APL Programs for the Mathematics Classroom

Norman Thomson

APL Programs for the Mathematics Classroom

Springer-Verlag
New York Berlin Heidelberg
London Paris Tokyo

Norman D. Thomson
IBM (United Kingdom) Laboratories Ltd.
Hursley Park
Winchester, Hampshire
England SO21 2JN

Library of Congress Cataloging-in-Publication Data
Thomson, Norman (Norman D.)
  APL programs for the mathematics classroom / Norman Thomson.
   p.   cm.
  Includes indexes.
  ISBN-13:978-0-387-97002-8   e-ISBN-13:978-1-4612-3668-9
  DOI: 10.1007/978-1-4612-3668-9

  1. Mathematics—Study and teaching—Computer programs. 2. APL
(Computer program language) I. Title.
  QA11.T625   1989
  510'.78—dc20   89-6263
Printed on acid-free paper.

© 1989 by Springer-Verlag New York, Inc.

All rights reserved. This work may not be translated or copied in whole or in part without the written permission of the publisher (Springer-Verlag, 175 Fifth Avenue, New York, NY 10010, USA), except for brief excerpts in connection with reviews or scholarly analysis. Use in connection with any form of information storage and retrieval, electronic adaptation, computer software, or by similar or dissimilar methodology now known or hereafter developed is forbidden.
The use of general descriptive names, trade names, trademarks, etc. in this publication, even if the former are not especially identified, is not to be taken as a sign that such names, as understood by the Trade Marks and Merchandise Marks Act, may accordingly be used freely by anyone.

Camera-ready text provided by author.

9 8 7 6 5 4 3 2 1

ISBN-13:978-0-387-97002-8   Springer-Verlag New York Berlin Heidelberg
ISBN-13:978-1-4612-3668-6   Springer-Verlag Berlin Heidelberg New York

Preface

The idea for this book grew out of proposals at the APL86 conference in Manchester which led to the initiation of the I-APL (International APL) project, and through it to the availability of an interpreter which would bring the advantages of APL within the means of vast numbers of school children and their teachers.

The motivation is that once school teachers have glimpsed the possibilities, there will be a place for an "ideas" book of short programs which will enable useful algorithms to be brought rapidly into classroom use, and perhaps even to be written and developed in front of the class.

A scan of the contents will show how the conciseness of APL makes it possible to address a huge range of topics in a small number of pages. There is naturally a degree of idiosyncrasy in the choice of topics - the selection I have made reflects algorithms which have either proved useful in real work, or which have caught my imagination as candidates for demonstrating the value of APL as a mathematical notation. Where appropriate, notes on the programs are intended to show the naturalness with which APL deals with the mathematics concerned, and to establish that APL is not, as is often supposed, an unreadable language written in a bizarre character set.

Of course, the _really_ important algorithms are the ones you program yourself, and the present collection is perhaps best considered as a stimulus rather than an authoritative reference.

At the time of writing, I-APL is available on the IBM PC and look alikes, and versions for other machines are imminent. In the spirit of the I-APL project, the proceeds of this book will be donated entirely to the furtherance of I-APL, and thereby to the promotion of APL as a tool for teaching. In so doing, it is good to recognise the tireless efforts of Anthony Camacho and Ed Cherlin in bringing the project to its present state of fruition, as well as the cooperation in I-APL publications of Linda Alvord, Tama Traberman and Garry Helzer, and the help and support of many other people throughout the world.

Norman Thomson
IBM (UK) Laboratories,
Hursley Park,
WINCHESTER,
Hants. England. SO21 2JN

Contents

Chapter 1. Introduction	1
1.1. Graphics	5
1.2. Idioms	6
Chapter 2. Arithmetic and Numbers	7
2.1. Basic Programs with Integers	7
2.2. Square and Triangular Numbers	8
2.3. Multiplication and other Tables	10
2.3.1. Log Tables	11
2.3.2. Trig Tables	11
2.4. Isomorphisms	11
2.5. Primes and Factors	12
2.6. HCF and LCM	15
2.7. Recurring Decimals	16
2.8. Numbers in Different Bases	16
2.9. Roman Numerals	18
2.10. Encoding and Decoding	19
2.11. Problems involving Base 10 Digits	21
2.12. Computer Arithmetic	22
2.13. Counting Series Forwards and Backwards	23
2.14 Complex Numbers	24
2.14.1. Complex Roots of Unity	25
Chapter 3. Algebra and Sets	27
3.1. Some Basic Algebra	27

3.2. Roots of Quadratics 29
3.3. Matrix Operations 30
    3.3.1. Determinants 31
3.4. Polynomials 31
3.5. Arithmetic and Geometric Progressions 33
3.6. Sets 34
3.7. Polynomial Coefficients from Roots 36

Chapter 4. Series 39
4.1. Recurrence Relations 39
4.2. Tests for Monotonicity 42
4.3. Convergence 42
4.4. Binomial Coefficients 43
    4.4.1. Pascal's Triangle 44
4.5. Successive Differences of Series 45
4.6. Fibonacci Numbers 46
4.7. Series relating to pi 47
4.8. Series for e 50
4.9. A Series for $\sqrt{2}$ 50
4.10. Trig Series 51
4.11. Continued Fractions 51
4.12. Interpolation 52

Chapter 5. Formulae and Tables 55
5.1. Compound Interest 56
    5.1.1. Present Values 58
5.2. Mortgage Repayments 58
5.3. Triangle Formulae 59
5.4. Longest and Shortest Journeys 61
5.5. Pythagoras's Theorem and Norms 62
5.6. Pythagorean Triples 63

Chapter 6. Geometry and Pattern 65
6.1. Parametric Plotting 67
    6.1.1. Conic Sections 67
    6.1.2. Hypocycloids and Epicycloids 68
6.2. Envelopes 71
    6.2.1. Conic Sections 72
    6.2.2. Hypocycloids and Epicycloids 75

Contents   IX

6.3. Transformations   77
6.4. Perspective Drawing   80
6.5. Co-ordinate Geometry in Two Dimensions   81
6.6. Polar and Cartesian Coordinates   82
6.7. Patterns by Plotting Large Numbers of Points   84

Chapter 7. Calculus   87
7.1. Numerical Integration   87
    7.1.1. Upper and Lower Bounds for Integration   88
    7.1.2. Trapezium Rule   89
    7.1.3. Simpson's Rule   89
    7.1.4. Adaptive Simpson's Rule   90
7.2. Root Finding   91
    7.2.1. Bisection Method   91
    7.2.2. Iteration Method   92
    7.2.3. Newton-Raphson Method   94
7.3. Ordinary Differential Equations   96
    7.3.1. Euler's Method   96
    7.3.2. Mid-point Method   97
    7.3.3. Trapezium Method   97

Chapter 8. Probability and Statistics   99
8.1. Discrete Probability Distributions   99
    8.1.1. Binomial Distribution   99
    8.1.2. Poisson Distribution   100
    8.1.3. Hypergeometric Distribution   101
8.2. The Birthday Problem   102
8.3. Descriptive Statistics   103
    8.3.1. Variance   103
    8.3.2. Standard Deviation   104
    8.3.3. Partition Values   104
    8.3.4. Mode and Range   105
8.4. Random Numbers from Various Distributions   106
8.5. Simulations   108
    8.5.1. Dice/Coins etc.   108
    8.5.2. Buffon's Needle   108
8.6. Frequency Distributions   109
    8.6.1. Stem-and-leaf Plot   110
    8.6.2. Two-way Frequency Distribution   111
    8.6.3. Scatterplots   112

8.7. Regression 113
8.8. Correlation 115
    8.8.1. Covariance and Correlation Matrices 115
8.9. Non-parametric Tests 117
    8.9.1. Runs Test 117
    8.9.2. Rank Correlation 118
        8.9.2.1. Spearman's Coefficient 118
        8.9.2.2. Kendall's Coefficient 119
    8.9.3. Sign Test 120
    8.9.4. Wilcoxon Signed Rank Test (W-test) 122
    8.9.5. Mann-Whitney Rank Sum Test (U-test) 123
    8.9.6. Goodness of Fit 125
8.10. Statistical Tables 126
    8.10.1. Normal probability density fn. 126
    8.10.2. Student t probability density fn. 128
    8.10.3. Gamma and Chi-squared proby. dens. fns. 129
    8.10.4. F probability density funtion 130
8.11. Sample Sizes 131

Chapter 9. Combinatorics 135
9.1. Permutations in Lexical Order 135
9.2. Derangements 138
9.3. Combinations 140
9.4. Selections 141
9.5. Compositions and Partitions 142
9.6. Latin Squares 143
9.7. Magic Squares of Odd Order 144

Chapter 10. Games and Miscellaneous 145
10.1. Deal a Hand at Whist 145
10.2. Chessboard 146
10.3. Mastermind 147
10.4. Life 148
10.5. Recursive Algorithms 149
    10.5.1. Tower of Hanoi 149
    10.5.2. Ackerman's Function 150
10.6. Optical Illusions 150

Contents

Appendix 1. Graphics	153
Appendix 2. Idioms and Utilities	161
A2.1. Rounding, Averaging, and Removing Duplicates	161
A2.2. Sorting and Ranking	162
A2.3. Statement Joining	165
A2.4. Branching and Prompting	167
A2.5. Matrix Manipulation	168
A2.6. Replication	169
A2.7. Without	170
A2.8. Bit Manipulation	170
A2.9. Some String Handling Functions	171
A2.10. Testing for Numeric/Character	173
A2.11. Timing Function Execution	174
Appendix 3. Graphics Functions in I-APL	175
Index of Topics	179
Index of Programs and Variables	183

Chapter 1

Introduction

This compendium is designed to provide mathematics teachers with a wide range of mathematical programs of relevance to current syllabuses. A further aim is to provide these in small bulk, the combination of breadth and conciseness being achieved through use of APL which allows many mathematical algorithms to be expressed in a fraction of the number of statements that are required in other languages, thereby allowing the programming language to be truly subservient to the mathematics.

This book is not designed to be in itself a tutorial for APL. It is assumed that the reader either possesses a reasonable practising knowledge of APL, or has studied one of the available tutorial introductions to the language.

There are four essential reasons why computers should appear at all in the Maths classroom/laboratory, namely Computation, Investigation, Display, and Simulation.

Some of the programs in this collection have the aim of describing an established algorithm or technique, thus fulfilling a computational role. Other programs are given with a view to suggesting an investigation. Here the algorithm is of no enduring value of itself - its value is rather in the mathematical insight to be gained from writing it and observing its execution. Such programs will be prefixed with the letter I for Investigation. Other programs are given which enhance mathematical insight by producing some numerical result in graphical form - these programs reference functions described in Appendix 1, which are external to APL and depend on a supporting graphics system. Examples are to be found in Chapters 4 and 6. Finally, there are programs which carry out some specific simulation, e.g. allocating birthdays or dealing a hand at whist. The aim has been to range in level from school to University mathematics, and it is hoped that a wide range of practising mathematics teachers who can get their hands on a computer with APL will find something of relevance in these pages.

While APL is not unique in providing these capabilities, it has the overwhelming advantage of being much closer to conventional mathematical notation in its constructs than any other programming language available on microcomputers. This substantially reduces the danger of the computing tail wagging the mathematical dog, and of what starts out to be an investigation in mathematics becoming a wrestle with the intricacies of a computer system. The computer should become an unobtrusive presence in the classroom, available to provide immediate support when the progress of instruction makes it appropriate. In this way APL promises to be the biggest single motivator ever to arrive in the mathematics classroom.

The term "program" will be used to cover several different constructs. In the first place many of the programs given are accommodated on a single line as "phrases." Sometimes phrases embody a condition, in which case they have 3 parts, the condition, the "action-if-true," and the "action-if-false." Other phrases are compounds of two simple phrases separated with a statement connector. There are thus three types of phrases, viz. simple phrases, conditional phrases, and compound phrases. All phrases are named for possible cross-reference elsewhere in the collection. Such names are generally function names in the APL sense, although when a phrase is a simple transliteration of a mathematical formula, as e.g. in the case of compound interest,

1. Introduction

its name is taken to be the variable on the receiving side of the assignment statement. When it is not possible to embody an algorithm in the form of a phrase, it is made the subject of a multi-line program.

Many of the multi-line functions follow a simple basic structure whose outline is

```
[1]      initialisation
[2]   L1:branch on condition
[3]      iterative action
[4]      →L1
```

In the interests of conciseness, the programs are devoid of the sort of data checks that are characteristic of commercial programs in which such "cosmetics" often occupy a large bulk of the program code. The aim here is rather to use the programming language in ways that highlight mathematics.

A few conventions are used to make for uniformity of style.

(a) To help illustrate transition from mathematics to APL, equations and expressions in this font will are in conventional notation, while the heavier font is always pure APL.

(b) In general the result of a program is denoted by Z, and the left and right arguments by L and R respectively. With function phrases however, the assignment(s) to Z are implicit, following the convention of direct definition form, which is described under (f) below.

(c) Where it is necessary to give explicit descriptions of the arguments L and R, these usually follow the statement of the function or phrase.

(d) Temporary variables are denoted by T,U,T1,T2,...; if these are counting variables I,J,...are used, and where context makes it desirable X,Y,N,H are also used as temporary variables.

(e) Labels are denoted by L1,L2,... and are sequenced in order of appearance.

(f) For phrases, a formulation known as "direct definition" is used. This is available in I-APL where α and ω are used to represent left and right arguments, rather than L and R. Where a program comprises a conditional statement together with two alternative statements one or other of which is executed according to the truth or falsity of the condition, this is presented in a single line in the form :

	statement		conditional		statement
name :	executed if	:	statement	:	executed if
	condition false				condition true

This should be read as :

"name"
       is defined as
"..."
       unless
"..."
       in which case
"..."

where the underlined words are the expansions of the colons.

(g) Sometimes, in order to express a program as a phrase, or to group together two short similar or connected statements, the symbol '&' is used as a statement connector. On some APL systems it is necessary to write such statements on two separate lines, the leftmost one first.

(h) Another form of statement compression combines a branch and an assignment on the same line, e.g.

$\rightarrow L1, Z \leftarrow \ldots$

1. Introduction

This should be read as "branch to L1, having set Z to ..." (See also Appendix 2 — Merging Branch Statements).

(i) Index origin is 1 throughout.

(j) Some algorithms are given in a recursive form, where this is judged to give insight into the underlying mathematics. In such cases the direct definition pattern under (f) above can usually be interpreted :

    name : recursive : stopping : stopping
            action   condition   action

As above, read the colons as "is defined as," "unless," and "in which case."

(k) The symbol $\leftrightarrow$ is used to denote "is equivalent to."

(l) The APL code used conforms to the ISO APL standard with the addition that $0 \top R$ is used to mean "round to the nearest integer."

1.1. Graphics

Graphics functions enable mathematical processes to be <u>watched</u> e.g. the convergence of a series or the solution of a differential equation. To exploit APL fully the teacher should make the extra effort required to realise such graphics using whatever ancillary package is available. Whereas the APL language itself is consistent over many interpreters, the techniques for producing graphics vary from one computer system to another. A few basic graphical facilities are all that is needed in order to generate useful displays. Appendix 1 contains a list of these described in a non-implementation-specific way, while Appendix 3 contains functions which implement them on IBM PC's and compatible machines using a workspace PGRAPH supplied with I-APL.

1.2. Idioms

There are some phrases which have occurred with great frequency in practical use of the APL language, so much so that experienced users immediately recognize them in the context of other programs. Some of these have little to do with mathematics, but are concerned with matters such as restructuring and displaying data — topics which cannot be altogether ignored in using APL for mathematics. Appendix 2 gives a list of idioms relevant within the scope of the present volume.

Chapter 2

Arithmetic and Numbers

2.1. Basic Programs with Integers

The index generator ι is a versatile tool for generating regular sets of integers, for example

First R positive integers	: ιR
First R even integers	: $2 \times \iota R$
First R multiples of L	: $L \times \iota R$

The first R odd integers are given by:

$$ODDS \quad : \quad ^{-}1+2\times\iota R$$

Example :

```
      ODDS 5
1 3 5 7 9
```

The technique embodied in the above program can be extended to any problem involving series of numbers spaced at regular intervals, and will be considered further under the heading "Arithmetic Progressions" in Chapter 3.

As a particular case consider the problem of scaling and labelling the axis of a graph. If the labelled points are to be from -20 to 25 in 9 steps of 5, think of them first as

$$(-20) + 0, 5, 10, \ldots 45$$

which is ¯20+5×0,ι9 in APL.

This leads to the program

```
    AXISLAB:R[1]+((-/R[2 1])÷R[3])×0,ιR[3]
```

> R : left-hand end of scale,
> right-hand end of scale,
> number of intervals

```
     AXISLAB ¯20 25 9
¯20 ¯15 ¯10 ¯5 0 5 10 15 20 25
```

2.2. Square and Triangular numbers

First R square numbers   : (ιR)*2 or +\ODDS R

First R triangular numbers : +\ιR or 2!-ιR
                             (i.e. as binomial coefficients)

APL can provide insight into <u>why</u> the first of these pairs of programs are equivalent. The underlying process can be generalised

2. Arithmetic and Numbers

in the following way to calculate the first R Lth. powers of integers, i.e. $(\iota R)\star L$.

```
        ∇ Z←L POWERS R;I
[1]     I←L  &  Z←ι1+L×R-1
[2]     L1:→0 IF 0=I←I-1
[3]     →L1,Z←+\((ρZ)ρ(Iρ1),0)/Z
        ∇
```

A trace of this function used to calculate the first 3 4th. powers is:

```
        4 POWERS 3
1 3 6 11 17 24 33 43
1 4 15 32 65 108
1 16 81
```

First we observe the cumulative sum of ι9 omitting every fourth item, then the cumulative sum of this result omitting every third item, and finally the cumulative sum omitting every second item.

I. Which numbers are both square and triangular? A suggested program which goes as far as the first R triangular numbers is

$$SQUTRI : (T\epsilon R\star 2)/T\leftarrow+\backslash R\leftarrow \iota R$$

I. Which integers have the property that the last 3 digits of their squares equal the number itself?

```
        ∇ Z←SQXXX R;I
[1]     I←R[1]  &  Z←ι0
[2]     L1:→0 IF R[2]<I←I+1
[3]     →L1,Z←Z,(I=1000|I*2)/I
        ∇
```

         R : start and end points of search

2.3. Multiplication and other Tables

Table generation can demonstrate the power of APL to good effect as in the following program for which R is the size of the required table.

$$MULTAB\ :\ R\circ.\times R\leftarrow\iota R$$

Example :

```
      MULTAB 6
 1  2  3  4  5  6
 2  4  6  8 10 12
 3  6  9 12 15 18
 4  8 12 16 20 24
 5 10 15 20 25 30
 6 12 18 24 30 36
```

Tables for addition, subtraction, exponentiation and other arithmetic functions are obtained similarly. For more on such tables see Chapter 5.

The next program uses the encode function to find the row and column number of the first occurrence of the value R in the table L, read in row-major order.

$$LOOKUP\ :\ 1+(\rho L)\top ^{-}1+(,L)\iota R$$

Example :

```
      (MULTAB 6)LOOKUP 12
2 6
```

2. Arithmetic and Numbers

2.3.1. Log Tables

A set of standard 4-figure log tables in base 10 is given by

$$LOG \leftarrow 6 \; 4 \top 10 \circledast (.1 \times 9 + \iota 90) \circ . + .01 \times {}^{-}1 + \iota 10$$

Anti-logarithms are :

$$ANTILOG \leftarrow 0 \top 10 * 3 + (.01 \times {}^{-}1 + \iota 100) \circ . + \\ .001 \times {}^{-}1 + \iota 10$$

2.3.2. Trigonometry Tables

An entire set of sin/cos/tan tables can be obtained by

$$1 \; 2 \; 3 \circ . O R$$

where R is a table of values. The table of values is in turn described by an outer product, e.g.

$$(0, \iota 90) \circ . + .1 \times 0, \iota 9$$

to give a typical set of values from 0 to 90 degrees by 0.1. Allowing for conversion to radians, the trig tables are given by

$$TRIGTAB \leftarrow 1 \; 2 \; 3 \circ . O (O \div 180) \times (0, \iota 90) \circ . + .1 \times 0, \iota 9$$

2.4. Isomorphisms

Isomorphisms between various addition and multiplication tables in "clock arithmetic" can readily be investigated using the outer product operator. For example, the addition table for integers modulo 4 has the same structure as the multiplication table for 1 2 4 3 modulo 4 and for 1 3 9 7 modulo 10. This equivalence is demonstrated by using these tables as indices to appropriate vectors, viz:

```
'ABCD'[1+4|T∘.+T←0 1 2 3]
'ABDC'[5|T∘.×T←1 2 4 3]
'A B  D C'[10|T∘.×T←1 3 9 7]
```

In all cases the result is

ABCD
BCDA
CDAB
DABC

Another isomorphism concerns multiplication modulo 8 and multiplication modulo 12:

```
'A B C D'[8|T∘.×T←1 3 5 7]
'A  B C  D'[12|T∘.×T←1 5 7 11]
```

In both cases the result is

ABCD
BADC
CDAB
DCBA

2.5. Primes and Factors

Primes up to R

The following phrase defines the primes up to R as those of the first R positive integers which have exactly two factors, namely 1 and the integer itself.

```
PRIMES : (2=+⌿0=R∘.|R)/R←⍳R
```

2. Arithmetic and Numbers   13

Here is an alternative program which expresses directly the fundamental property that primes are the numbers which do <u>not</u> appear anywhere in the body of the multiplication table.

$$PRIMES : (\sim R \in R \circ . \times R)/R \leftarrow 1 \downarrow \iota R$$

Example :

```
    PRIMES 30
2 3 5 7 11 13 17 19 23 29
```

Factors of R (including R itself)

The following phrase selects those integers which divide a given integer R exactly.

$$FACTORS: (0=(\iota R)|R)/\iota R$$

Example :

```
    FACTORS 12
1 2 3 4 6 12
```

Factors of R (excluding R itself)

This phrase uses the fact that no factor of R can exceed ½R.

$$FACS \quad : (0=T|R)/T \leftarrow \iota \lfloor .5 \times R$$

Example :

```
    FACS 12
1 2 3 4 6
```

Problems involving Factors

I. What can be said about integers with an even number of factors ? Here is a program which prints such integers in the range 1 to R:

$$EVENFAC:(EVENFAC\ R-1), R \times \iota 0 = 2 \mid \rho FACS\ R:$$
$$R=0\ :\ \iota 0$$

The same recursive program structure is used in the next example.

I. Which are the "perfect" numbers, i.e. numbers which are equal to the sum of their factors ? The perfect numbers in the range L to R are given by:

$$PERFNOS:(L\ PERFNOS\ R-1), R \times \iota R = +/FACS\ R:$$
$$R=L\ :\ \iota 0$$

A faster way to generate perfect numbers is to observe that $(2*R-1) \times {}^-1+2*R$ is perfect provided that both R and ${}^-1+2*R$ are prime.

I. Investigate $+/\div FACTORS\ R$ for R both perfect and non-perfect.

I. Find some pairs of "amicable" numbers, i.e. numbers such that each is the sum of factors of the other. Sometimes with problems like this it is sensible to leave the computer running indefinitely to report "interesting" cases as it discovers them, with the final program stop occurring on user interrupt. Here is a suggested program in this style.

2. Arithmetic and Numbers   15

```
      ∇ AMICNOS;I
[1]   I←0
[2]   L1:→L1 IF (+/FACS +/FACS I)≠I←I+1
[3]   →L1 IF I=+/FACS I
[4]   I,+/FACS I
[5]   →L1
      ∇
```

I. Are there numbers which form a 3-link chain? simply add another `+/FACS` to line 2, and find out!

2.6. HCF and LCM

Highest Common Factor

A simple way to calculate hcf(L,R) is to find the largest factor of L which is also a factor of R :

$$HCF\ :\ {}^{-}1\uparrow(T\epsilon FACTORS\ R)/T\leftarrow FACTORS\ L$$

Another algorithm for obtaining the HCF of two numbers is Euclid's algorithm, which is recursive in structure:

$$EUC\ :\ L\ EUC\ (L\leftarrow L\lfloor R)|L\lceil R\ :\ 0\epsilon L,R\ :\ L\lceil R$$

The basis of the algorithm is that if T is the smaller the two integers L and R, then

$$hcf(L,R)\ =\ hcf(T,\ max(L,R)(mod\ T))\ .$$

This process of reduction to simpler integer pairs continues until one of the integers is 0, in which case the hcf is the other.

Lowest Common Multiple

$$LCM \; : \; (L \times R) \div L \; HCF \; R$$

Examples :

```
      12 EUC 18
6
      12 LCM 18
36
```

2.7. Recurring decimals

The following program obtains as a vector the cycle of digits which recur when the non-terminating decimal fraction L/R (L, R integer) is expanded.

$$RECUR \; : \; L \; RECUR1 \; L,R$$
$$RECUR1 \; : \; (\lfloor T \div R[2]),U \; RECUR1 \; R \; :$$
$$R[1]=U \leftarrow R[2]|T \leftarrow 10 \times L \; : \; \lfloor T \div R[2]$$

Examples :

```
          2 RECUR 7
2 8 5 7 1 4
          1 RECUR 17
0 5 8 8 2 3 5 2 9 4 1 1 7 6 4 7
```

2.8. Numbers in Different Bases

The following series of problems involve conversions between numbers in decimal notation and their representations in other number bases.

Express a positive integer R (or a vector of positive integers R) as a minimum length vector of digits in number base L :

2. Arithmetic and Numbers

```
       BASE : ((⌊1+L⍟1⌈/,R)⍴L)⊤R
```

Examples :

```
       8 BASE 21
2 5
       10 BASE 23 7
2 0
3 7
```

Express vector R as row vectors of digits of length $L[1]$ in number base $L[2]$:

```
       EXPRESS : ⍉(L[1]⍴L[2])⊤R
```

Evaluate a matrix R of row vectors interpreted as numbers in base L :

```
       EVAL : L⊥⍉R
```

Examples :

```
       3 8 EXPRESS 21 100
0 2 5
1 4 4
       8 EVAL 3 8 EXPRESS 21 100
21 100
```

i.e. 21 is 25_8, and 100 is 144_8.

Express decimal number R as a whole number part followed by first L digits of binary fraction :

```
       BINFRAC :(1↑T),(L-1)BINFRAC 2×
              1↓T←,0 1⊤R : L=¯1 : ⍳0
```

Express result of BINFRAC as a decimal number :

    DECFRAC : R+.×2*-¯1+ιρR

Examples :

      5 BINFRAC 7.7
7 1 0 1 1 0

      DECFRAC 7 1 0 1 1 0
7.6875

2.9. Roman Numerals

Here are procedures for dealing with another number base!

Convert Roman Numeral to Integer

      ∇ Z←ARABIC R;T
[1]   T←1000 500 100 50 10 5 1['MDCLXVI'ιR]
[2]   Z←—+/T×¯1*T≥(ρT)↑1↓T
      ∇

             R : Character string representing a Roman
                 numeral

2. Arithmetic and Numbers

Convert Integer R to Roman

```
      ∇ Z←ROMAN R;I;T;U
[1]    T←'MDCLXVI' & U←0 2 5 2 5 2 5⊤R
[2]    Z←U[1]ρ'M' & I←1
[3]    L1:→0 IF 7<I←I+1
[4]    →L2 IF U[I]=4
[5]    Z←Z,U[I]ρT[I]
[6]    →L1
[7]    L2:→L3 IF U[I-1]≠0
[8]    Z←Z,T[I-0 1]
[9]    →L1
[10]   L3:Z←(¯1↓Z),T[I-0 2]
[11]   →L1
      ∇
```

Examples :

```
      ARABIC 'MCMLXXVIII'
1978
      ROMAN 1978
MCMLXXVIII
```

2.10. Encoding and Decoding

The functions ⊤ and ⊥ are amazingly versatile. 1⊥R for example is equivalent to +/R, and 0⊥R is equivalent to selection along the last axis of R. Here are some further applications of these two primitive functions.

Separate R into its integral and fractional parts :

$$SEP \; : \; 0 \; 1\top R$$

R : scalar, array or vector

Example :

        SEP 4.76 150
4   150
0.76   0

Money Conversion

How many 10p., 5p., and 1p. coins are required to make up a sum of R pence using the minimum number of coins?

        $COINS$: 0 2 5⊤R

Example :

        $COINS$ 77
7 1 2

Mensuration : Conversion of Units

Convert a vector R representing yards, feet and inches into a vector representing the same length in m., cm., and mm. :

        $METRIC$: 0 100 10⊤25.4×0 3 12⊥R

Example :

        $METRIC$ 3 2 6
3 50 5.2

that is, 3 yards 2 feet 6 inches is equivalent to 3m. 50 cm. 5.2mm.

2. Arithmetic and Numbers   21

2.11. Problems involving Base 10 Digits

Find the sum of the digits in a base 10 integer R, (or vector of base 10 integers) :

    $DIGSUM : +/10\ BASE\ R$

Example :

    $DIGSUM\ 23\ 199$
5 19

I. The following program returns 1 if the two arguments L and R contain exactly the same digits (possibly in a different order).

    $SAMEDIG: \wedge/(T \in U), (U \leftarrow ,10\ BASE\ R) \in$
                                         $T \leftarrow ,10\ BASE\ L$

Use SAMEDIG to find all the pairs of 2-digit integers such as (21,87) whose product 1827 consists of the same 4 digits.

```
      ∇ RPT;I;J
[1]   I←10
[2]   L1:→0 IF 31<I←I+1
[3]   J←⌊1000÷I
[4]   L2:→L1 IF 99<J←J+1
[5]   →L2 IF ~(I,J)SAMEDIG I×J
[6]   I,J
[7]   →L2
      ∇
```

I. Casting out Nines

The following program describes a number game and uses the function ASK in Appendix 2. Type CAST9, and follow the instructions -

```
       ∇ CAST9;T
[1]    T←ASK'GIVE A NUMBER WITH ITS DIGITS
                              ALL DIFFERENT:'
[2]    T←T,ASK'..AND ANOTHER USING THE SAME
                                    DIGITS:'
[3]    'DIGIT SUM OF LARGER - SMALLER IS ',
                                  ⍕DIGSUM |-/T
       ∇
```

— after a while you may prefer to speed up play by using the simulated form CAST.

```
CAST: DIGSUM |-/10⊥T[R?R],[1.5]
             T←¯1+R?10
```

R : integer between 2 and 10

2.12. Computer Arithmetic

It is interesting to work out for one's own computer the maximum precision with which a number is represented. One simple way to do this is to find the maximum value of N for which

$$Z←2*N \quad \& \quad Z←Z-1$$

results in a value of Z which is not 1. Since $L*R$ is likely to be evaluated within the machine as $*R×⍟L$, this test is valid only if you trust that the exponentiation and log routines are implemented using the full possible precision. If not, here is a program you can use which calculates the successive powers of 2 (line 3) stopping whenever the difference between $2*T$ and $(2*T)-1$ is not computed as 1.

2. Arithmetic and Numbers

```
       ∇ PRECTST;I;T
[1]    I←1 &  T←2
[2]    L1:→L2 IF 1≠T-T-1
[3]    I←I+1 &  T←2×T
[4]    →L1
[5]    L2:'MAX PRECISION IS ',(⍕I-1),' BITS'
       ∇
```

Example (on one specific computer) :

```
       PRECTST
MAX PRECISION IS 56 BITS
```

2.13. Counting Series Forwards and Backwards

Another comparison which demonstrates that the arithmetical precision on any computer is finite involves taking a series which has a large number of very small terms in its tail, and observing the difference between counting the terms starting from the head and then from the tail. From a numerical point of view it is better to cumulate from the tail, since if the big terms have been added first, the addition of small terms, however many, may have no effect at all on the floating point sum. If the small terms are added first, however, their cumulative effect will make a contribution. The following two programs do forward and backward summation for the sums of R terms of the series whose Ith. term is $I\star-L$. In general they will give different results, but you may have to look at a very low precision digit to find the difference!

```
       ∇ Z←L FWD R;I
[1]    I←Z←0
[2]    L1:→0 IF R<I←I+1
[3]    →L1,Z←Z+I⋆-L
       ∇
```

```
        ∇ Z←L BWD R;I
[1]     I←R+1  & Z←0
[2]     L1:→0 IF 0=I←I-1
[3]     →L1,Z←Z+I*-L
        ∇
```

You should also observe which, if either, of these sums is the same as

$$+/(\iota R)*-L$$

Example (on one specific computer) :

```
      3 FWD 100
1.20200740065967
      3 BWD 100
1.20200740065968
      +/(⍳100)*¯3
1.20200740065968
```

2.14. Complex Numbers

On some APL systems complex numbers are available as simple scalars. Where this is not the case the complex number a + ib may be represented by the two-element vector (a,b). Addition and subtraction are given by APL vector addition and subtraction; multiplication and division are given by

```
    MUL :  (-/L×R),+/L×⌽R

    DIV :  L⌹2 2⍴R,R MUL 0 1
```

Examples :

```
      2 1 MUL 3 ¯4
10 ¯5
```

2. Arithmetic and Numbers 25

```
        10 ¯5 DIV 2 1
3 ¯4
```

For argument and amplitude, see section 6.6 - conversion to and from polar and cartesian co-ordinates.

2.14.1. Complex Roots of Unity

The complex Rth. roots of 1 are given by:

$$CRU : \varphi(2,R)\rho(2\circ T), 1\circ T \leftarrow ((\circ 2) \div R) \times {}^{-}1 \times \iota R$$

Example :

```
       3⍕CRU 3
¯0.500  ¯0.866
¯0.500   0.866
 1.000   0.000
```

Chapter 3

Algebra and Sets

3.1. Some Basic Algebra

This section begins with some programs designed to support some of the earliest algebraic ideas introduced in schools. These programs illustrate the use of the format function (⍕) to mix calculations and conversation, and also a prompting function ASK which is given in Appendix 2.

Difference of two squares

```
      ∇ DIFFSQ;T
[1]   T←ASK'GIVE 2 NOS.:'
[2]   'DIFF OF SQQ = ',(⍕T[1]*2),
         '-',(⍕T[2]*2),'=',⍕-/T*2
[3]   '(X+Y)×(X-Y) = ',(⍕+/T),'×',
         (⍕-/T),'=',⍕×/(+/T),-/T
      ∇
```

Square of sum and sum of squares

```
        ∇ SUMSQQ;T
[1]     T←ASK'GIVE 2 NOS.:'
[2]     'SQ. OF SUM = ',(⍕+/T),'*2 = ',
                                        ⍕(+/T)*2
[3]     'SUM OF SQQ = ',(⍕T[1]*2),'+',
                              (⍕T[2]*2),'=',⍕+/T*2
[4]     'DIFFERENCE = 2×',(⍕T[1]),'×',
                              (⍕T[2]),'=',⍕2××/T
        ∇
```

Examples :

```
      DIFFSQ
GIVE 2 NOS.:
4 3
DIFF OF SQQ = 16-9=7
(X+Y)×(X-Y) = 7×1=7

      SUMSQQ
GIVE 2 NOS.:
4 3
SQ. OF SUM = 7*2 = 49
SUM OF SQQ = 16+9=25
DIFFERENCE = 2×4×3=24
```

Expansion of Brackets

The next function concerns the expansion of a pair of terms each of the form (ax + by)(cx + dy). R is a 2x2 matrix the first row of which is the value of the coefficients a,b and the second row is the coefficients c,d. The result is the coefficients of the expansion in the order x^2, xy, y^2.

```
      EXPAND : ((×/R),+/×/0 1⌽R)[1 3 2]
```

3. Algebra and Sets

Example :

        EXPAND 2 2ρ1 2 3 ¯4
3 2 ¯8

3.2. Roots of Quadratics

This is written in the conditional direct definition form, with the 3 sections in order corresponding to (1) real roots, (2) $b^2-4ac > 0$ test, (3) imaginary roots.

        QUAD: (−R[2]+1 ¯1×T⋆.5)÷2×R[1] :
                 0>T←¯2 1 ¯2+.×R×ΦR :
    'IMAG. ROOTS:',⍕((−R[2]),(|T)⋆.5)÷2×R[1]

        R : vector of coefficients in descending power order

The expression ¯2 1 ¯2+.×R×ΦR is an APL version of b^2-4ac.

Where the roots are imaginary, the result is given as a 2-item vector a,b which is to be interpreted as (a+ib), (a-ib).

Example :

        QUAD 1 2 1
¯1 ¯1
        QUAD 1 1 1
IMAG. ROOTS:¯0.5 0.866

3.3. Matrix Operations

Product of two (compatible) matrices L and R : $L+.\times R$

Inverse of square matrix R : $⌹R$

Quotient of two (compatible) matrices L and R : $L⌹R$

Solution of simultaneous linear equations : $L⌹R$

where the equations to be solved are Rx = L. If L is a matrix with columns l_1, l_2, ..., $L⌹R$ is the <u>set</u> of solutions Rx = l_1, Rx = l_2, etc.

Example :

```
        R              L
  1  1  1          4   6
  2 ¯1 ¯1          2  ¯9
  1  2  1          3   9
        L⌹R
   2 ¯1
  ¯1  3
   3  4
```

i.e. solution of   x + y + z = 4   is x = 2, y = -1, z = 3 ;
                  2x - y - z = 2
                   x + 2y + z = 3

− change RHS to 6 -9 9, solution is x = -1, y = 3, z = 4.

I. Confirm the following :

(i) If A = BC in the matrix multiplication sense, then $(⌹B)+.\times A$ and $A+.\times ⌹C$ are equal to C and B respectively.

3. Algebra and Sets

(ii) $A+.\times \boxminus A$ and $(\boxminus A)+.\times A$ are both equal to $UNIT$ $1\uparrow\rho A$
(See Appendix 2 for UNIT).

3.3.1. Determinants

The determinant of a square matrix R is given by the following algorithm which obtains the result as the product of the successive topmost leading diagonal items which appear as the original matrix is reduced by progressively replacing it with the outer product of its leading row and column divided by the current topmost diagonal element.

```
      ∇ Z←DET R;I
[1]   Z←R[1;1]
[2]   L1:Z←Z×1 1↑R←(1 1↓R)-((1↓R[;1])
                     ∘.×1↓R[1;])÷R[1;1]
[3]   →L1 IF 1<1↑ρR
      ∇
```

Examples :

```
      DET 4 4ρι16
0
      DET 2 2ρ5 2 7 3
1
```

3.4. Polynomials

In the examples in this section a polynomial is represented by a vector of its coefficients in descending power order. Coefficients of omitted powers must be represented by 0's.

Evaluate polynomial R at value(s) L

$$POLY\ :\ (\lozenge((\rho R),\rho L)\rho L)\bot R$$

Example :

```
      1 2 POLY 2 1 3
6 13
```

i.e. values of $2x^2 + x + 3$ at $x = 1$ and $x = 2$ are 6 and 13.

Addition and multiplication of two polynomials R and L

$$PADD\ :\ (-(\rho L)\lceil \rho R)\uparrow((L\neq L),R)+(R\neq R),L$$

$$PMULT\ :\ 1\downarrow +/\lozenge(-\iota\rho L)\phi L\circ.\times R,L\neq L$$

Examples :

```
      2 1 3 PADD 3 0 2 1
3 2 3 4
      2 1 3 PMULT 3 0 2 1
6 3 13 4 7 3
```

i.e. $(2x^2 + x + 3) + (3x^3 + 2x + 1) = 3x^3 + 2x^2 + 3x + 4$ and

$(2x^2 + x + 3)(3x^3 + 2x + 1) = 6x^5 + 3x^4 + 13x^3 + 4x^2 + 7x + 3$.

Differentiation and integration of polynomial R

$$PDIFF\ :\ ^{-}1\downarrow R\times\phi^{-}1+\iota\rho R$$

$$PINTEG\ :\ (R\div\phi\iota\rho R),0$$

Examples :

3. Algebra and Sets 33

```
      PDIFF 2 1 3
4 1
      PINTEG 2 1 3
0.667 0.5 3 0
```

Best Fitting Polynomial of given degree

    - see POLYFIT in Chapter 8

3.5. Arithmetic and Geometric Progressions

The formula for an arithmetic progression is:

    AP : $R[1]+R[2]\times 0,\iota R[3]-1$

        R : initial value (a),
           difference (d),
           no of terms (n)

and its sum is $+/AP$.

Example :

```
      AP 1 2 6
1 3 5 7 9 11
      +/AP 1 2 6
36
```

The formula for a geometric progression has the same structure, and differs only in the arithmetical functions involved:

    GP : $R[1]\times R[2]*0,\iota R[3]-1$

        R : initial value (a),
           ratio (r),
           no of terms (n)

and its sum is +/GP .

Example :

```
      GP 1 2 6
1 2 4 8 16 32
      +/GP 1 2 6
63
```

I. Confirm that the sums as given above conform to the standard mathematical formulae, namely

$$S = 2a + (n-1).d \quad \text{(AP)}$$

$$S = \frac{a(r^n - 1)}{(r - 1)} \quad \text{(GP)}$$

3.6. Sets

A set is conveniently modelled as a vector (numeric or character) with no repeated elements. The APL function ρ then gives the number of elements in the set, and membership of a set can be tested for by the function ϵ. It is also convenient to use the functions **WITHOUT** and **REMDUP** which are defined in Appendix 2.

Union and intersection of two sets L and R are then given by

```
    UN    : REMDUP L,R

    IX    : L WITHOUT L WITHOUT R
```

The best way in which to use the computer to illustrate the concepts of sets is to define a few simple vectors, e.g.

```
    U←ι10
    A←ι5
```

3. Algebra and Sets

```
B←2+ι6
C←2×ι4
```

and verify the various rules associated with sets, e.g.

Distribution properties of union and intersection

```
A UN B IX C  ↔  (A UN B) IX (A UN C)

A IX B UN C  ↔  (A IX B) UN (A IX C)
```

De Moivre's Laws

```
U WITHOUT A IX B  ↔
        (U WITHOUT A) UN (U WITHOUT B)
U WITHOUT A UN B  ↔
        (U WITHOUT A) IX (U WITHOUT B)
```

Subsets and supersets

If L is a subset of R, then the following subset and superset relations are true :

```
SUBSET  :  ∧/L∈R

SUPSET  :  R SUBSET L
```

All subsets of a set

```
ALLS    :  T⊤ι×/T←Rρ2
```

Example :

```
    ALLS 3
0 0 0 1 1 1 1 0
0 1 1 0 0 1 1 0
1 0 1 0 1 0 1 0
```

An alternative recursive version which brings the last column to the front is :

$$ALLS:(0,[1]T),1,[1]$$
$$T \leftarrow ALLS\ R-1\ :\ R=1\ :\ 1\ 2\rho 0\ 1$$

All subsets of size L

$$SOMESET\ :\ (L=+\neq T)/T \leftarrow ALLS\ R$$

Example :

```
      2 SOMESET 3
0 1 1
1 0 1
1 1 0
```

This program can be used to find for a 3-item vector **R**

(a) sums of all pairs of items   : $R+.\times 2\ SOMESET\ 3$

(b) products of all pairs of items   : $R\times.*2\ SOMESET\ 3$

3.7. Polynomial Coefficients from Roots

The following function gives the coefficients of in descending power order of the polynomial whose roots are given by the items of vector **R**.

$$PCOFS\ :\ (^{-}1*^{-}1+\iota\rho Z)\times Z \leftarrow ((0,\iota\rho R)\circ.=+\neq T)$$
$$+.\times R\times.*T \leftarrow ALLS\ \rho R$$

Examples : Coefficients of $(x-1)^3$ and $(x-1)(x-2)(x-3)$.

```
     PCOFS 1 1 1
1 ¯3 3 ¯1
```

3. Algebra and Sets

```
      PCOFS 1 2 3
1 ¯6 11 ¯6
```

This function demonstrates the use of inner and outer products to good advantage.

First the columns of ALLS select all possible sets of roots. Next the ×.⋆ inner product converts each such set to a product of roots (note that $N \star 0 = 1$ for all N)

The expression $((0, \iota \rho R) \circ . = + \neq T)$ gives a classification of these products with each row corresponding to the number of components $(0,1,...)$. The +.× inner product then sums the products of roots according to these groupings.

Finally $(^{-}1 \star ^{-}1 + \iota \rho Z)$ adjusts for the fact that the sums of the products of roots yield the polynomial coefficients with alternating signs.

Chapter 4

Series

4.1. Recurrence Relations

Terms of first order recurrence relation.

Two programs are given, the first constructs the first L terms of the series $u_n = k_0 + k_1 u_{n-1}$, and the second prints all the terms using a recursive algorithm.

```
        ∇ Z←L SER1 R;I
[1]     Z←R[,I←1]
[2]     L1:→0 IF L<I←I+1
[3]     →L1,Z←Z,(¯2↑R)+.×1,¯1↑Z
        ∇
```

              L : number of terms required
              R : u_0, k_0, k_1

   $TERM1$: L $TERM1(\Box\leftarrow R)\bot\phi L$

L : k_0, k_1
R : u_0

Example :

```
      8 SER1 1 1 2
1 3 7 15 31 63 127 255
      1 2 TERM1 1
1
3
7
15
31    etc.
```

Terms of second and higher order recurrence relations.

The analogous functions SER2 and TERM2 deal with the second order recurrence relation $u_n = k_0 + k_1 u_{n-1} + k_2 u_{n-2}$ with initial values u_0, u_1.

```
     ∇ Z←L SER2 R;I
[1]  Z←(I←2)↑R
[2]  L1:→0 IF L<I←I+1
[3]  →L1,Z←Z,(¯3↑R)+.×1,¯2↑Z
     ∇
```

L : number of terms required
R : u_0, u_1, k_0, k_1, k_2

TERM2 : L TERM2(☐←¯1↑R),L+.×1,R

L : k_0, k_1, k_2
R : u_0, u_1

Example: The first 8 terms of the Fibonacci series starting with 0 1 are thus

4. Series

```
        8 SER2 0 1 0 1 1
0 1 1 2 3 5 8 13
```

and the terms of the Fibonacci series are generated by

```
    0 1 1 TERM2 0 1
```

The generalisations of SER2 and TERM2 to higher order relations are:

```
      ∇ Z←L SER R;I;T;U
[1]   Z←(I←⌊.5×ρR)↑R & T←I
[2]   U←I↓R
[3]   L1:→0 IF L<I←I+1
[4]   →L1,Z←Z,U+.×1,T↑Z
      ∇
```

      L : number of terms required
      R : $u_0, u_1,..., k_0, k_1, k_2,...$

    TERM : ```L TERM(□←¯1↑R),(1↓¯1↓R),L+.×1,R```

      L : k's in ascending subscript order
      R : u's in ascending subscript order

I. Find the periodicity of the series $u_n = abs(u_{n-1}) - u_{n-2}$. The following program produces the series for different start values.

```
      ∇ Z←L SERX R;I
[1]   Z←R & I←2
[2]   L1:→0 IF L<I←I+1
[3]   →L1,Z←Z,(|¯1↑Z)-Z[¯1+ρZ]
      ∇
```

      L : number of terms required
      R : u_0, u_1

4.2. Tests for Monotonicity

The following phrases return 1 for a strictly decreasing sequence, 0 otherwise.

$$MONOD \ : \ \wedge/{}^{-}1{\downarrow}R>1{\phi}R \quad \text{or} \quad MONOD \ : \ R<.\leq 1{\phi}R$$

The following phrases return 1 for a strictly increasing sequence, 0 otherwise.

$$MONOI \ : \ \wedge/{}^{-}1{\downarrow}R<1{\phi}R \quad \text{or} \quad MONOI \ : \ R<.\geq 1{\phi}R$$

Note that the first versions of MONOD and MONOI could equivalently be expressed as inner products :

$$MONOD \ : \ ({}^{-}1{\downarrow}R)\wedge.>1{\downarrow}R$$

$$MONOI \ : \ ({}^{-}1{\downarrow}R)\wedge.<1{\downarrow}R$$

If "strictly decreasing" and "strictly increasing" are relaxed to "non-increasing" and "non-decreasing" the > and < in the above should be changed to ≥ and ≤ respectively.

4.3. Convergence

The ι function is valuable in helping to judge whether an infinite series is or is not convergent. The general principle is to define $T \leftarrow \iota 25$ say, and then do a + scan on an expression which defines the series. Here are a few simple examples with the conventional mathematical form for u_n in brackets following :

$$+\backslash \div T \quad\quad\quad (\ u_n = 1/n \ )$$
$$+\backslash \div T\star 2 \quad\quad (\ u_n = 1/n^2 \ )$$
$$+\backslash \div !T \quad\quad\quad (\ u_n = 1/n! \ )$$
$$+\backslash \div T\times T+1 \quad (\ u_n = 1/n(n+1) \ )$$

4. Series

```
+\(÷T)+⁻1*T      (  uₙ = (1/n)+(-1)ⁿ  )
+\÷1↓⍟T          (  uₙ = 1/ln n  )
```

Applying the DRAW function (see Appendix 1) with T as x-axis then gives a vivid realisation of the behaviour of the series. For example

```
0 0 25 2 DRAW
    T,+\÷(T*2),(!T),[1.5]T×T+1
```

gives (from top to bottom) plots of $\Sigma 1/n!$, $\Sigma 1/n^2$, and $\Sigma 1/n(n+1)$.

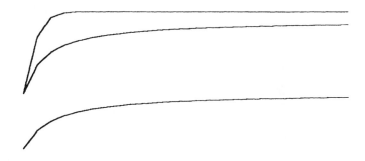

4.4. Binomial Coefficients

First L coefficients of expansion of $(x+y)^R$. R may be integral or non-integral, positive or negative.

```
BINCOEF:  (0,ιL-1)!R
```

Examples :

```
      4 BINCOEF 4
1 4 6 4
      4 BINCOEF ¯2
1 ¯2 3 ¯4
      4 BINCOEF ¯3
1 ¯3 6 ¯10          (triangular nos. with alternating signs)
      4 BINCOEF ¯.5
1 ¯0.5 0.375 ¯0.3125
```

4.4.1. Pascal's Triangle

The coefficients of the terms in the expansion of $(1+x)^R$ for integers up to R are given as columns of the matrix defined by

$$PAS \; : \; R\circ.!R\leftarrow 0,\iota R$$

For coefficients of $(1-x)^R$ use

$$PASN \; : \; \phi R\circ.!R\leftarrow \iota R+1$$

Examples :

```
      ⌈PAS 4                    ⌈PASN 4
1 1 1 1 1                0 0 0 0 1
0 1 2 3 4                0 0 0 1 ¯1
0 0 1 3 6                0 0 1 ¯2 1
0 0 0 1 4                0 1 ¯3 3 ¯1
0 0 0 0 1                1 ¯4 6 ¯4 1
```

(The ⌈'s are present to force the result to integer, and thereby make the output display more concise.)

4. Series

4.5. Successive differences of series

The following function returns a vector of length $(\rho R)-L$ resulting from taking differences L times.

$DIF : (1\downarrow T)-{}^{-}1\downarrow T\leftarrow(L-1)DIF\ R\ :\ L=0\ :\ R$

    L : number of differences to be taken
    R : series in vector form

If a table of successive differences is required, insert $\square\leftarrow$ before $T\leftarrow(L-1)DIF\ R$.

Example :

```
      2 DIF (ι7)*2
1  4  9 16 25 36 49
3 5 7 9 11 13
2 2 2 2 2
```

I. Using the variation suggested above, investigate the following successive differences :

    $4\ DIF\ |8\ BINCOEF\ {}^{-}4$

    $5\ DIF\ |8\ BINCOEF\ {}^{-}5$

Difference tables are an excellent means of detecting isolated errors in series, as the following investigation demonstrates.

I. Type the following sequence, and observe the effect of the artificially introduced error.
    $T\leftarrow|20\ BINCOEF\ {}^{-}5$
    $0\rho5\ DIF\ T$
    $T[7]\leftarrow T[7]+1$
    $0\rho5\ DIF\ T$

4.6. Fibonacci Numbers

Fibonacci numbers satisfy the recurrence relationship $u_n = u_{n-1} + u_{n-2}$. Such series are generated by the following iterative program :

```
      ∇ Z←L FIB R;I
[1]   I←2 & Z←R
[2]   L1:→0 IF L<I←I+1
[3]   →L1,Z←Z,+/¯2↑Z
      ∇
```

                L : number of terms (≥ 2)
                R : u_0, u_1

Example :

```
      10 FIB 0 1
0 1 1 2 3 5 8 13 21 34
```

Add `[2.1]` `÷/¯2↑Z` to print the ratios of successive terms.

```
      8 FIB 0 1
0
1
0.5
0.667
0.6
0.625
0 1 1 2 3 5 8 13
```

A recursive function to output the terms of the series is

        `FIB1 : R FIB1 R+⎕←L`

where L and R are u_0 and u_1 respectively. Adding a stopping condition allows a clean finish, e.g.

4. Series

$FIB2$: R $FIB2$ $R+\square \leftarrow L$: $L \geq 1000$: L

I. (i) Define $T \leftarrow 15$ FIB 0 1 and compare $(2 \downarrow T) \times \bar{\ } 2 \downarrow T$ with $1 \downarrow \bar{\ } 1 \downarrow T \star 2$.

(ii) Define $T \leftarrow 30$ FIB 1 1 and compare $T[M]$ HCF $T[N]$ and $T[M \; HCF \; N]$ for pairs of integers M and N between 1 and 30.

(iii) Define

$T \leftarrow 60$ FIB 1 1

and investigate $T[N] | T[N \times \iota \lfloor 60 \div N]$ for values of N in the range 3 to 10. Describe the property of the Fibonacci series which this shows.

(iv) What is the result of $(N \; FIB \; 0 \; 1)+.\times PAS \; N-1$ for an appropriate small integer value of N ?

4.7. Series relating to pi

In each case R is the number of terms, and the function defines a series which converges to π.

Gregory's Series $\pi/4 = 1 - 3^{-1} + 5^{-1} - ...$

$GREGORY$: $4 \times -\backslash \div ODDS \; R$

Euler's Series $\pi/4 = (2^{-1} + 3^{-1}) - (2^{-3} + 3^{-3})/3 + (2^{-5} + 3^{-5})/5 - ...$

$EULER$: $4 \times -\backslash (+/(\div 3 \; 2)\circ.\star R) \div R \leftarrow ODDS \; R$

Sum of reciprocals of squares of integers

$SREC2$: $(6 \times +\backslash \div (\iota R)\star 2)\star .5$

Sum of reciprocals of fourth powers of integers

    SREC4 : (90×+\÷(⍳R)*4)*.25

I. Find the constants analagous to 6 and 90 which occur in the similarly defined functions SREC6, SREC8, SREC10, SREC12.

It is useful to show the convergence behaviour of these series by

    0 2.5 25 4 DRAW (⍳25),(GREGORY 25),
         (EULER 25),(SREC2 25),[1.5]SREC4 25

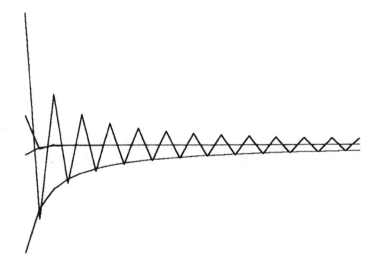

which graphs them in the following top to bottom order of their leftmost points − GREGORY, EULER, SREC4, SREC2.

Further series which converge to multiples of π

Product of $(1 - 1/p^2)$ for all primes p:

    PI1 : 6××\1-÷(PRIMES R)*2

Sums of reciprocal squares excluding multiples of 3:

4. Series

$$PI2 : (6.75×+\÷((0≠3|R)/R←\iota R)*2)*.5$$

Sums of reciprocal squares of form 6n+1 or 6n-1:

$$PI3 : (9×+\÷(((6|R+1)\in 0\ 2)/R←\iota R)*2)*.5$$

Sum of reciprocals of squares of integers with alternating signs:

$$PI4 : (12×+\(R\rho 1\ ^-1)×÷(\iota R)*2)*.5$$

Sum of quotients of factorials divided by products of odd numbers:

$$PI5 : +\2×(!0,\iota R)÷×\ODDS\ R+1$$

Examples :

```
      7 3⊤PI1 30
 4.500   4.000   3.840   3.762   3.731   3.708
                 3.696   3.685   3.678   3.674
      7 3⊤PI2 15
 2.598   2.905   2.976   3.021   3.044   3.061
                 3.072   3.082   3.088   3.094
      7 3⊤PI3 30
 3.000   3.059   3.089   3.101   3.110   3.115
                 3.119   3.122   3.124   3.126
      7 3⊤PI4 10
 3.464   3.000   3.215   3.096   3.172   3.119
                 3.158   3.128   3.152   3.133
      7 3⊤PI5 10
 2.667   2.933   3.048   3.098   3.122   3.132
                 3.137   3.139   3.141   3.141
```

4.8. Series for e

(a) Rapidly convergent $E1$: `+\÷!0,⍳R`

(b) Slowly convergent $E2$: `(1+÷⍳R)*⍳R`

Examples :

```
      E1 6
1 2 2.5 2.6667 2.7083 2.7167 2.7181 2.7183
      E2 10
2 2.25 2.3704 2.4414 2.4883 2.5216 2.5465
               2.5658 2.5812 2.5937 2.6042
```

4.9. A Series for the square root of 2

The following program gives the values of the fractions 1/1, 3/2, 7/5, 17/12, ... in which each is derived from its predecessor m/n as (m + 2n)/(m + n):

```
      ∇ Z←L ROOT2 R;I
[1]     I←0  &  Z←÷/R
[2]     L1:→0 IF L≤I←I+1
[3]     R←(1 2+.×R),+/R
[4]     →L1,Z←Z,÷/R
      ∇
```

               L : number of terms
               R : two initial terms

Example :

```
      8 ROOT2 1 1
1 1.5 1.4 1.4167 1.4138 1.4143 1.4142 1.4142
```

4. Series   51

4.10. Trig Series

The successive series expansions of the first L terms for cosR and sinR are :

$$COS \ : \ -\backslash(R\star T)\div!T\leftarrow 2\times{}^{-}1+\iota L$$

$$SIN \ : \ -\backslash(R\star T)\div!T\leftarrow ODDS \ L$$

These can be compared with the values of $2oL$ and $1oL$.

Examples :

```
        (5 COS 1),2o1
1 0.5 0.54167 0.54028 0.5403 0.5403
        (5 SIN 1),1o1
1 0.83333 0.84167 0.84147 0.84147 0.84147
```

4.11. Continued Fractions

Evaluate the continued fraction corresponding to vector R

$$CONFRAC: \ R[1]+\div R[2] \ : \ 2<\rho R \ :$$
$$R[1]+\div CONFRAC \ 1\downarrow R$$

The simplest continued fraction is $1+1/(1+1/(1+1/(...$ which converges to $\frac{1}{2}(1+\sqrt{5}) = 1.618$. Other interesting limits are given in the table below in which $T+1$ is the length of the vector generating the continued fraction.

CONFRAC $1,T\rho 2$	$\sqrt{2}$
CONFRAC $1,T\rho 1 \ 2$	$\sqrt{3}$
CONFRAC $2,T\rho 4$	$\sqrt{5}$
CONFRAC $2,T\rho 2 \ 4$	$\sqrt{6}$
CONFRAC $2,T\rho 1 \ 4$	$\sqrt{8}$
CONFRAC $3,T\rho 6$	$\sqrt{10}$
CONFRAC $2,T\rho 3 \ 6$	$\sqrt{11}$

```
    CONFRAC 2,Tρ2 6      √12
    CONFRAC 2,Tρ1 6      √15
    CONFRAC 4,Tρ8        √17
    CONFRAC 4,Tρ4 8      √18
    CONFRAC 4,Tρ1 8      √24
```

In general

```
    CONFRAC N,Tρ2×N      √(1+N²)
```

To observe the convergence, insert ⎕← before $R[1]$ in the rightmost conditional expression.

Example :

```
    CONFRAC 1,6ρ2
2.4
2.4167
2.4138
2.4143
1.4142
1.4142
```

4.12. Interpolation

If a function f has $y_0 = f(x_0)$ and $y_1 = f(x_1)$ the linearly interpolated estimate for f(x) is given by the formula

$$\frac{y_0 (x - x_1)}{(x_0 - x_1)} + \frac{y_1 (x - x_0)}{(x_1 - x_0)}.$$

A more general form of this is called Lagrange's formula, which for a table R of (n + 1) columns

$$\begin{array}{cccc} y_0 & y_1 & \cdots & y_n \\ x_0 & x_1 & \cdots & x_n \end{array}$$

4. Series

uses the best fitting polynomial of degree n to interpolate y at a specified x-value L.

```
        Z←L LAGR R;T;I
[1]     Z←I←0
[2]     L1:→0 IF(¯1↑ρR)<I←I+1
[3]     T←R[2;(ι¯1↑ρR)WITHOUT I]
[4]     Z←Z+R[1;I]×÷/×/(L,R[2;I])∘.-T
[5]     →L1
     ∇
```

$\quad$ WITHOUT : (~L∈R)/L

Example : Find the 5th. degree Lagrangian interpolation for the value x = π/6 in the table T which gives values of sin x:

```
     T←(10T),[.5]T←.1×ι6
     T
0.09983 0.1987 0.2955 0.3894 0.4794 0.5646
0.1     0.2    0.3    0.4    0.5    0.6
     (○÷6)LAGR T
0.5
```

Linear interpolation between the two closest points gives

```
     (○÷6)LAGR 0 4↓T
0.4995
```

Another technique for polynomial interpolation is to obtain vectors of linear interpolations on the polynomial interpolations of successively higher degree, each time offsetting the x-values (but not the y-values) by one further position. This technique is known as Neville's algorithm. The best estimate at any stage is that for which the the tabular x-values are closest to and on either side of the x-value for interpolation. L is the polynomial degree and R is a table such as T with (0,x) appended as first column.

NEV : $(-/(\ominus 1\ OFF(L-1)NEV\ R) \times T-R[2;1])$
    $\div -/T \leftarrow L\ OFF\ 1\downarrow R[2;]$: $L=0$: $1\downarrow R[1;]$

OFF : $(0,-L)\downarrow (L\phi R),[.5]R$

Example : Find the 2nd degree polynomial interpolation for $x = \pi/6$ in the table T above.

```
      T1←(0,○÷6),T
      2 NEV T1
0.5049 0.5014 0.5001 0.5
```

In practice, it is more useful to have an algorithm which carries on increasing the degree of the interpolation until two successive estimates agree within a specified tolerance. In POLI below, L is the required tolerance, R is defined as for NEV. The first line does a column reordering so that the first two x-values are closest to and on either side of the x-value for interpolation. If there are insufficient tabulated values to achieve the given tolerance, the result of POLI is a null vector.

```
        ∇ Z←L POLI R;T;I;X
[1]     R←R[;1+0,▲|R[2;1]-1↓R[2;]]
[2]     X←1↓R[2;]
[3]     I←0,0ρZ←1↓R[1;]
[4]     L1:→(L<|-/(1↑Z←(-/(⊖1 OFF Z)×T-R[2;1])
                ÷-/T←(I←I+1)OFF X),1↑Z)/L1
[5]     Z←(0≠ρZ)↑Z
        ∇
```

Example : Interpolate table T at x = $\pi/6$ with tolerance .001.

```
      .001 POLI T1
0.5
```

Chapter 5

Formulae and Tables

In using APL to do straightforward evaluations of arithmetic formulae, the input expression is usually so close to the algebraic statement of the formula that it scarcely merits formal description as a program. One of the considerable advantages of APL over other programming languages is that where only one algebraic quantity is involved in the formula this can often be generalized to a vector, or indeed a higher order array, to give multiple evaluations with no change in APL input. Here are some examples:

Area of a circle   ○$R \star 2$

Volume of a sphere   ○$4 \times (R \star 3) \div 3$

Conversion from Centigrade to Fahrenheit $32+1.8 \times C$

If conversion of a range of Centigrade degrees to Fahrenheit were required at 10 degree intervals, use

$\qquad 32+1.8 \times C \leftarrow 0, 10 \times \iota 10$

When more than one variable is involved in the formula, outer products can be used to obtain simultaneous evaluations.

Examples :

Volume of a cylinder :	Formula	○$H \times R \star 2$
	Table	○$H \circ . \times R \star 2$
Surface area of a cone :	Formula	○$R \times L$
	Table	○$R \circ . \times L$
Compound Interest :	Formula	$A \times (1+R) \star N$
	Table	$A \circ . \times (1+R) \circ . \star N$
Velocity at time T :	Formula	$U + A \times T$
	Table	$U \circ . + A \circ . \times T$

5.1. Compound Interest

In the following, A stands for Amount, R for Rate, and N for Number of years. The problem is to compute the value of A following investment for N years at rate R. First assign values to A, N and R:

```
A←10000 20000 30000
R←.09 .10 .12
N←5 10 15 20
```

5. Formulae and Tables

```
      7 0⊤CI←A∘.×(1+R)∘.*N
15386   23674   36425    56044
16105   25937   41772    67275
17623   31058   54736    96463

30772   47347   72850   112088
32210   51875   83545   134550
35247   62117  109471   192926

46159   71021  109274   168132
48315   77812  125317   201825
52870   93175  164207   289389
```

The above formula can be generalised by making it into a function with A,R,N as right argument, and a vector $(\rho A),(\rho R),\rho N$ as left argument:

$$CI\ :(L[1]\uparrow R)\circ.\times(1+L[2]\uparrow L[1]\downarrow R)\circ.*(-L[3])\uparrow R$$

```
      7 0⊤3 3 4 CI 10000 20000 30000 .09 .10 .12
                                      5  10  15  20
15386   23674   36425    56044
16105   25937   41772    67275
17623   31058   54736    96463

30772   47347   72850   112088
32210   51875   83545   134550
35247   62117  109471   192926

46159   71021  109274   168132
48315   77812  125317   201825
52870   93175  164207   289389
```

5.1.1. Present Values

To calculate present values it is necessary only to change "multiply" in the above formula/function to "divide":

```
      7 0⌽PV←A∘.÷(1+R)∘.*N
   6499   4224   2745   1784
   6209   3855   2394   1486
   5674   3220   1827   1037

  12999   8448   5491   3569
  12418   7711   4788   2973
  11349   6439   3654   2073

  19498  12672   8236   5353
  18628  11566   7182   4459
  17023   9659   5481   3110
```

5.2. Mortgage Repayments

The problem is to compute repayments for loan A repaid over N years at rate R with interest compounded annually (values of A, R and N as above). The formula for the annual repayments in conventional notation is:

$$\frac{A.r.(1+r)^n}{(1+r)^n - 1}$$

Judicious insertion of outer products again gives the table from the APL formula.

Formula $(\div 12) \times A \times (R \times N \times T) \div {}^{-}1 + T \leftarrow (1+R)*N$
Table $MORT \leftarrow (\div 12) \times A \circ . \times ((R \circ . \times N * 0) \times T) \div {}^{-}1 +$
$T \leftarrow (1+R) \circ . * N$

5. Formulae and Tables

```
        7 2⍕MORT
214.24 129.85 103.38  91.29
219.83 135.62 109.56  97.88
231.17 147.49 122.35 111.57

428.49 259.70 206.76 182.58
439.66 271.24 219.12 195.77
462.35 294.97 244.71 223.13

642.73 389.55 310.15 273.87
659.49 406.86 328.68 293.65
693.52 442.46 367.06 334.70
```

Thus, for example, it costs 91.29 per month to repay 10000 at 9% over a period of 20 years. Notice how the "multiply" outer product balances array sizes rather than doing any direct computation.

5.3. Triangle Formulae

Area of Triangle given sides

The formula $\sqrt{s(s-a)(s-b)(s-c)}$ is given in APL as

       $TAREA$: $(\times/(.5\times+/R)-R,0)*.5$

              R : vector of lengths of 3 sides

Example :

       $TAREA$ 3 4 5

Sin Formula

If R is a 3-item vector representing the angles A and B in degrees, and the length of side b opposite B, then the length of side a is given by

$$SIDEA\ :\ R[3]\times\div/1\ 0\ 2\uparrow R$$

Example :

$$SIDEA\ 60\ 60\ 3$$
3

If R is a 3-item vector representing the sides a and b and the angle B opposite b in degrees, then one of the possible values for angle A (in degrees) is

$$ANGLEA\ :\ (180\div o1)\times{}^{-}1o(1ooR[3]\div180)$$
$$\times\div/2\uparrow R$$

Example :

$$ANGLEA\ 3\ 5\ 90$$
36.87

If R defines an impossible triangle, e.g. if a.sinB > b, then a DOMAIN ERROR will be reported for attempting the arcsine of a value greater than 1.

Cos Formula

If R is a 3-item vector representing the lengths of 2 sides and the angle in degrees between them, then the length of the third side is given by

$$SIDE3\ :\ ((T,2ooR[3]\div180)+.\times$$
$$T,\times/{}^{-}2,T\leftarrow2\uparrow R)*.5$$

5. Formulae and Tables   61

Example :

   $SIDE$ 3 3 4 90
5

5.4. Longest and Shortest Journeys

Given a distance table M for 4 towns, e.g.

```
        M
0  4  6  2
4  0  3  1
6  3  0  9
2  1  6  0
```

the longest 2-leg journeys which can be made between pairs of towns is

```
      M⌈.+M
12   9   8  15
 9   8  10  12
11  10  15   9
12   9   8  15
```

To find the shortest 2-leg journeys it is necessary to fill the leading diagonal of M with large numbers, e.g. 100 would do in this case:

```
         M
100    4    6    2
  4  100    3    1
  6    3  100    9
  2    1    6  100
```

```
        ML.+M
 4   3   7   5
 3   2   7   6
 7  10   6   4
 5   6   4   2
```

For journeys of more than 2 legs the inner product is repeated the appropriate number of times by using "execute" to generate the appropriate number of ⌈'s or ⌊'s.

       LONGEST: ⍎'R',(4×L−1)ρ'⌈.+R'

       SHRTEST: ⍎'R',(4×L−1)ρ'⌊.+R'

            L : no. of legs in journey
            R : distance table with large value in leading diagonal in case of SHRTEST

5.5. Pythagoras's Theorem and Norms

The formula $\sqrt{(a^2 + b^2 + c^2 + \ldots)}$ is given by:

       PY:(+/R*2)*.5
          R : vector a,b,c,...

Examples:

       PY 3 4
5
       PY 3 4 5
7.071

If R is generalised to be a matrix, each column of which is a vector of co-ordinates in n-dimensional space, an "outer product" of Euclidean norms is given by:

5. Formulae and Tables

```
        ∇ Z←NORM R;T
[1]     Z←1 1⌽T←(⌽R)+.×R
[2]     Z←((Z∘.+Z)−2×T)*.5
        ∇
```

Example :

```
      T
0  3 12
0  4  5
      NORM T
 0  5       13
 5  0        9.055
13  9.055    0
```

Another norm is sometimes called the "taxi-cab metric" or L-norm. It is defined as $|x_1 - x_0| + |y_1 - y_0|$ for points (x_0, y_0), (x_1, y_1). The corresponding outer product form is:

$$LNORM : ((⌽R)+.\lceil R)-(⌽R)+.\lfloor R$$

Example :

```
      LNORM T
 0   7 17
 7   0 10
17  10  0
```

5.6. Pythagorean Triples

Pythagorean triples are sets of integers which satisfy the equation $a^2 = b^2 + c^2$. The following function returns a single triple from scalar parameters L and R. These should be coprime to give a fundamental triple, i.e. one with no common factor amongst its terms.

$$TRIPLE : T[\triangle T←(2×L×R),|-/T],+/T←(L,R)*2$$

The next function generates families of triples — integer R is the stopping criterion; the number of triples is about (but not greater than) ½R(R + 1). For function EUC see section 2.6.

```
       ∇ Z←PYTH R;T;U
[1]    T←1
[2]    L1:→0 IF (1+2×R)<T←T+2
[3]    U←¯1
[4]    L2:→L1 IF T<U←U+2
[5]    →L2 IF 1≠U EUC T-U
[6]    →L2,□←U TRIPLE T-U
       ∇
```

Example :

```
       PYTH 4
3  4  5
8  15 17
5  12 13
12 35 37
7  24 25
20 21 29
16 63 65
9  40 41
28 45 53
```

Chapter 6

Geometry and Pattern

Most of the programs in this chapter compute arrays of numerical values representing configurations of lines and points, which are supplied as right arguments to the graphics programs described in Appendix 1 to produce geometrical displays on the screen. The functions which are used in this chapter are summarised here with diagrams to provide a visual reminder of the parameters.

(a) DRAW   : window DRAW (NxP) matrix

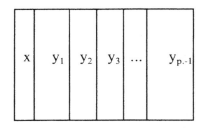

and similarly for PLOT except that points are not joined.

(b) JOIN   : window JOIN (x,y)

This is a "draw-as-you-go" function, which is used to interleave calculating and plotting, e.g. when there is not enough workspace to store an entire array for DRAW.

(c) LINES : window LINES (Nx4) matrix

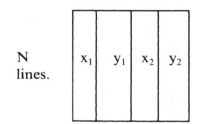

N lines.

(d) SKETCH : window SKETCH (Nx3) matrix

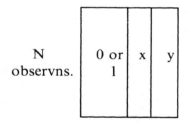

N observns.

Column 1 = 0 for pen-up, 1 for pen-down.

(e) CIRCLES : window CIRCLES (Nx3) matrix

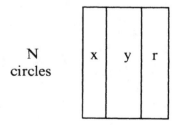

N circles

6. Geometry and Pattern

6.1. Parametric Plotting

The points of a circle of radius L are defined by the single parameter θ where

$x = L\cos\theta$
$y = L\sin\theta$.

This leads to the following function whose result is a 2-column matrix whose rows are coordinates of points on the circumference of a circle.

      CIRC : ⌽*L*×2 1∘.○○*R*÷180

           L : radius
           R : vector of angles in degrees

Further examples of parametric plotting follow.

6.1.1. Conic Sections

Matrix of points on ellipse and hyperbola

      ELLIPSE:((ρ*Z*)ρ*L*)×*Z*←1 *CIRC R*

           L : 2-item vector — major, minor semi-axes
           R : vector of angles in degrees

      HYPERB : *REFLX*((ρ*Z*)ρ*L*)×
                 Z←⌽6 5∘.○○*R*÷180

           L : 2-item vector — major, minor semi-axes
           R : vector of angles in degrees for
              one half of curve

$REFLX$: $(\ominus R),[1]R\times(\rho R)\rho 1\ ^-1$

The auxiliary function **REFLX** returns the result of adding a reflection in the x-axis together with a reordering of the points for a continuous plot. To draw the complete hyperbola use

$^-14\ ^-7\ 14\ 7\ DRAW\ 2\ 1\ HYPERB\ 5\times\iota 30$

for the right hand half, then

$^-14\ ^-7\ 14\ 7\ DRAW\ ^-2\ 1\ HYPERB\ 5\times\iota 30$

for the left hand half.

Matrix of points on parabola

$PARAB$: $REFLX\ L\times(R*2),[1.5]2\times R$

L : parameter a in $x = at^2$, $y = 2at$
R : vector of values of parameter t

Example : The parabola $y^2 = 8x$ is drawn by

$0\ ^-50\ 300\ 50\ DRAW\ 2\ PARAB\ 0,\iota 12$

6.1.2. Hypocycloids and Epicycloids

These curves give widely varying, and in many cases aesthetically pleasing symmetric patterns constructed by joining successive points whose co-ordinates are given by the parametric equations:

x = cos t + a.cos bt
y = sin t + a.sin bt

```
    ∇ Z←L EHPARM R
[1] L←○(2×0,⍳L)÷L←|L×R[2]−1
[2] Z←((2○L)+R[1]×2○R[2]×L),[1.5]
              (1○L)+R[1]×1○R[2]×L
    ∇
```

6. Geometry and Pattern

L : (number of arcs)/abs(b-1)
R : parameters a and b

Example :

⁻4 ⁻4 4 4 *DRAW* 1.75 *EHPARM* 2 33

The above curve connects 56 points in 7 clusters of 8. Changing the parameters gives considerable scope for investigation and experimentation.

Hypocycloids and epicycloids are generated by a small circle rolling along the circumference of a larger circle − externally for an epicycloid, internally for a hypocycloid. The following alternative function gives the parameters a more obvious physical meaning :

EHCURVE: *R*[2] *EHPARM* (*R*[3]÷1−*R*[3]),1+*R*[1]

R : circles per revolution,
   no. of points plotted on each circle,
   ratio of radius of revolving circle to
   that of base circle(< 1)

R[1]>0 gives an epicycloid, R[1]<0 gives a hypocycloid. R[2] need not be an integer, but the product ×/R[1 2] must be integral.

Example :

¯4 ¯4 4 4 *DRAW EHCURVE* ¯64 2.9375 .65

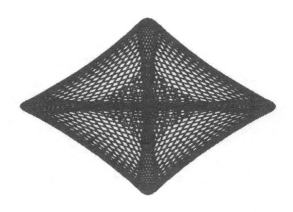

Again, changing the parameters gives considerable variation in the patterns produced.

6. Geometry and Pattern

6.2. Envelopes

An envelope is a curve which touches each member of a system of lines or curves. One of the simplest such systems is the "sliding ladder" curve or "astroid" :

```
      ∇ Z←ASTROID;I
[1]   Z←0 4ρI←.1
[2]   L1:→L2 IF 1<I←I+.1
[3]   Z←Z,[1]I,0 0,(1-I*2)*.5
[4]   →L1
[5]   L2:Z←4 REFLECT 1 REFLECT Z
      ∇
```

(see Appendix 1 for REFLECT)

Lines 1-4 describe the lines in the first quadrant using Pythagoras theorem (line 3) applied to a ladder of unit length. Line 5 produces the reflections in the other three quadrants. The picture is produced by e.g.

¯1.5 ¯1.5 1.5 1.5 *LINES ASTROID*

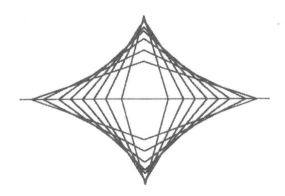

6.2.1. Conic Sections

Ellipses and hyperbolae

Start with a base circle (say the unit circle) and a point (focus) on the x-axis. Join the focus to a point on the circle and draw a line at right angles — this line is one element of the envelope.

```
        ∇ Z←L CONIC R;T
[1]     L←1 CIRC(360÷L)×0,ιL
[2]     Z←(L[;1]-(L[;2]×2+L[;2])÷T←R-L[;1]),
                                        [1.5]¯2
[3]     Z←Z,(L[;1]+(L[;2]×2-L[;2])÷T),[1.5]2
        ∇
                L : no. of lines in envelope
                R : x co-ordinate of focus
```

6. Geometry and Pattern

If the absolute value of the right argument R of CONIC is less than 1 the result is an ellipse, if greater than 1 it is a half of a hyperbola. An ellipse with foci at (.1,0) and (-.1,0) is given by

    ¯3 ¯3 3 3 *LINES* 35 *CONIC* .1

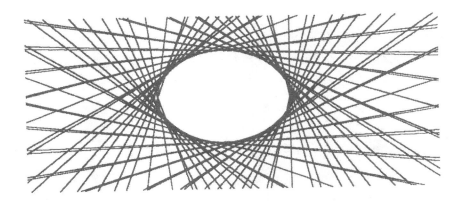

The right half of a hyperbola is given by

    ¯3 ¯3 3 3 *LINES* 35 *CONIC* 1.1

The second half is given by changing 1.1 to -1.1 so that the complete envelope is

    ¯3 ¯3 3 3 *LINES*
       (35 *CONIC* 1.1),[1]35 *CONIC* ¯1.1

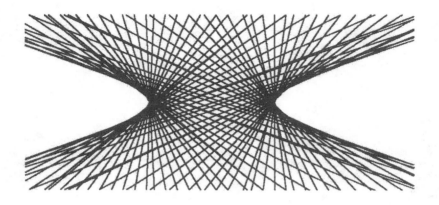

R = 1 gives a set of rays through the focus.

Avoid using "regular" values for both L and R (e.g. L = 36, R = .5) as this is likely to give a DOMAIN ERROR for division by 0 in line 1.

Parabola

A parabolic envelope is drawn by choosing a focus on the x-axis, and drawing rays through it to points on the y-axis. The lines at right angles to these rays form the required envelope.

```
PARENV : REFLX 0,(R×T),((4-R×T)×
            T←300(ιL)÷3×L),[1.5]4
```

   L : no. of lines in each half of envelope
   R : distance of focus from y-axis

Example :

   ¯1 ¯5 5 5 *LINES* 10 *PARENV* 1

6.2.2. Hypocycloids and Epicycloids

The following program generates an epicycloid if R is an integer greater than 1, and a hypocycloid if R is an integer less than -1. The number of cusps is abs(R-1).

```
     ∇ Z←L EPIHYP R;I;T;U
[1]    T←1 CIRC(360÷L)×0,⍳L
[2]    Z←0 4ρI←0
[3]    L1:→0 IF L<I←I+1
[4]    Z←Z,[1]T[I;],T[(L×U=0)+U←L|R×I;]
[5]    →L1
     ∇
```

These envelopes are generated by taking L equally spaced points on the circumference of a unit circle and joining each point I to the point RxI(mod L). Then use e.g.

¯2 ¯2 2 2 *LINES* 36 *EPIHYP* 2

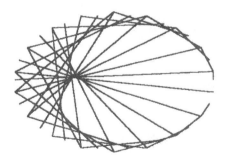

R = 2 produces the one-cusped epicycloid shown above, otherwise known as the cardioid. L = 36 is adequate for this. R = 3 produces the two-cusped epicycloid, sometimes known as the nephoid — L = 72 is suggested. Use L = 144 for R = 4, and

L = 216 for R = 6 and above. Higher values of L will give better resolution but may generate WS FULL.

To draw hypocycloids effectively, the line elements must be drawn <u>outside</u> the base circle. The function EXTEND replaces each chord with extensions of length L outwards from the base circle in each direction.

```
        ∇ Z←L EXTEND R;T
[1]     T←(-/R[;3 1]),[1.5]-/R[;4 2]
[2]     Z←R[;1 2],R[;1 2]-L×T
[3]     Z←Z,[1](R[;3 4]+L×T),R[;3 4]
        ∇
```

L = 2 for EXTEND, and L = 72 for EPIHYP will, together with a window ¯6 ¯6 6 6, give good representations of a wide range of hypocycloids, e.g.
¯6 ¯6 6 6 *LINES* 2 *EXTEND* 72 *EPIHYP* ¯2 gives a deltoid:

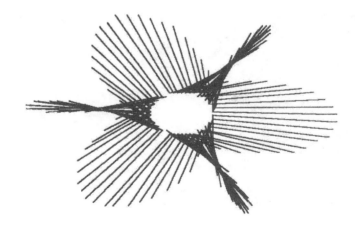

The cardioid can also be produced as a special case of a curve called a "limacon" which in turn may be drawn as an envelope of circles. The element of the envelope is a circle whose centre is a point on the base circle, and which goes through a fixed point.

6. Geometry and Pattern

The fixed point may be taken along the x-axis without loss of generality, and so the co-ordinates to be passed to CIRCLES are

$$LIMACON:R,(+/((-(\rho R)\rho L,0)+$$
$$R\leftarrow 1\ CIRC(360\div R)\times 0,\iota R)*2)*.5$$

        L : x-co-ordinate of fixed point
        R : no. of points on base circle

The special case where L is equal to the radius of the base circle is the cardioid. Take the window as (-(L+2), -(L+2), L+2, L+2). A typical picture is

      ¯3 ¯3 3 3 *CIRCLES* 1 *LIMACON* 18 (for cardioid)

Negative values of the left argument of LIMACON put the cusp on the other side; a value of 0 gives a figure of eight curve with a "knot" in the middle.

6.3. Transformations

Rotation about the origin in 2 dimensions

$$ROTO\ :\ R+.\times \mathbb{Q}2\ 1\circ.\circ(\circ L\div 180)+\circ 0\ 0.5$$

        L : anti-clockwise angle in degrees
        R : 2 element vector, or
            2 column matrix, each row of which is the
            (x,y) co-ordinate pair of a point

Rotation about a point other than the origin in 2-D

$$ROT2\ :\ T+L[3]\ ROTO\ R-T\leftarrow(\rho R)\rho L[1\ 2]$$

        L : x coord of point of rotation,
           y coord of point of rotation,
           anti-clockwise angle in degrees

R : 2 element vector, or
2 column matrix, each row of which is the
(x,y) co-ordinate pair of a point

Example : Sketch the triangle (1,2),(3,4),(4,3) and its image under rotation of 90 degrees about (1,1).

```
      ¯4 0 4 4 SKETCH
      (0 1 1 1, 1 1 90 ROT2 TRI),[1]
         0 1 1 1, TRI←4 2ρ1 2 3 4 4 3
```

Rotation in 3 dimensions

```
      ∇ Z←L ROT3 R
[1]   Z←R
[2]   Z[;2 3]←L[1] ROTO Z[;2 3]
[3]   Z[;1 3]←(-L[2]) ROTO Z[;1 3]
[4]   Z[;1 2]←L[3] ROTO Z[;1 2]
      ∇
```

L : anti-clockwise angle in degrees about Oz,
anti-clockwise angle in degrees about Oy,
anti-clockwise angle in degrees about Ox
R : 3 column matrix, each row of which is the
(x,y,z) co-ordinate triple of a point in 3D
space

One way to draw a projection of a cube is therefore to do a 3-dimensional rotation using ROT3 and then perform a DRAW on two of the resulting columns, which is equivalent to taking a section by one of the three co-ordinate planes.

If we take a unit cube with one its vertices at the origin, and the diagonally opposite vertex as (1,1,1), an edge trace is given by the matrix

6. Geometry and Pattern

```
CUBE← 16 3ρ 0 0 0
              0 0 1
              0 1 1
              0 1 0
              0 0 0
              1 0 0
              1 0 1
              0 0 1
              0 1 1
              1 1 1
              1 1 0
              1 0 0
              1 0 1
              1 1 1
              1 1 0
              0 1 0
```

and a projection on the Oyz plane is given by

```
¯2 ¯2 2 2 DRAW
    (40 50 20 ROT3 CUBE)[;2 3]
```

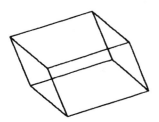

6.4. Perspective Drawing

```
PERSPEC: (R[;1]×T),[1.5]R[;2]×
              T←L÷L-R[;3]
```

L : perspective ratio (scalar) defined by the
mapping (x,y,z) to (xL/(L-z), yL/(L-z))
R : 3 column matrix, each row of which is the
(x,y,z) co-ordinate triple of a point in 3D
space

Taking CUBE as defined above, a perspective drawing is obtained by

```
¯4 ¯4 4 4 DRAW
    2 PERSPEC 40 50 20 ROT3 CUBE
```

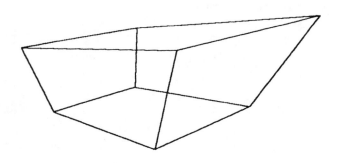

6. Geometry and Pattern

6.5. Co-ordinate Geometry in Two Dimensions

Argument of a vector

The following phrase returns $\sqrt{(l^2 + m^2)}$ where l and m are the 2 items of vector R.

$$ARG : (+/R\star 2)\star.5$$

Signed perpendicular distance of point from line lx + my + n = 0

Formula is $(lx_1 + my_1 + n)/\sqrt{(l^2 + m^2)}$. If $m \neq 0$ (i.e. the line is not parallel to the y-axis) distance is positive if the point is above the line, negative if below.

$$DIST : ((R,1)+.\times L) \div (\times L[2]) \times ARG\ 2 \uparrow L$$

> L : coefficients of line equation in form
> lx + my + n = 0
> R : 2 element vector (x,y), or
>   2 column matrix, each row of which is the
>   (x,y) co-ordinate pair of a point

If m = 0 the function is $R[1]+\div/L[3\ 1]$ with distance positive if the point is to the right of the line, negative if to the left. These two phrases could usefully be combined into a single function.

Image of point(s) in a line lx + my + n = 0 (l non-zero)

$$IMAGE : R-2\times(|L\ DIST\ R)\circ.\times$$
$$2\ 10^-30\div/L[2\ 1]$$

> L : coefficients l,m,n of equation of line
> R : 2 element vector (x,y), or
>   2 column matrix, each row of which is the
>   (x,y) co-ordinate pair of a point

If l = 0 then the appropriate formula is $R-2\times0,L$ $DIST$ R .

Angle between two lines

$$ANG \ : \ {}^{-}2{\circ}(R+.\times L)\div\times/(ARG\ R\leftarrow 2\uparrow R),$$
$$ARG\ L\leftarrow 2\uparrow L$$

returns the angle between two directed lines L and R each defined as a 3-item vector of coefficents of a line equation in the form lx + my + n = 0. Lines are determined by the ratios l, m and n rather than by their absolute values, and thus there are infinitely many representations of L and R. If the user is indifferent to which representation is used the result of ANG will be one of two angles which add up to π. ANG can however be interpreted more sensitively by observing a convention that if m≠0 then with conventional x and y axes, the positive direction of a line is upward or to the right if m > 0, and downward or to the left if m < 0. If m = 0 the direction is downward if l > 0. ANG then returns the angle between the positive directions of the lines L and R.

6.6. Polar and Cartesian Coordinates

$$CTOP \ : \ \lozenge((,\times 1\ 0/R)\times ARG\ R),[.5]$$
$$(180\div\circ 1)\times{}^{-}3{\circ}\div/\phi R$$

R : 2 element vector (x,y), or
2 column matrix, each row of which is
the (x,y) co-ordinate pair of a point

Example :

```
      S
1 1
⁻2 1
      CTOP S
```

6. Geometry and Pattern

```
 1.414    45
¯2.236  ¯26.57
```

$$PTOC \; : \; \mathbb{Q}((\rho T)\rho,1 \; 0/R)\times$$
$$T\leftarrow 2 \; 1\circ.\circ\circ(,0 \; 1/R)\div 180$$

> R : 2 element vector (R,θ) or
>   2 column matrix, each row of which is
>   the (R,θ) co-ordinate pair of a point

For all **X** it is true that

$$X \; \leftrightarrow \; PTOC \; CTOP \; X \; \leftrightarrow \; CTOP \; PTOC \; X$$

CTOP delivers its angle in an amplitude range of -90 to 90 degrees, and consequently, the value of the argument may be either positive or negative. If a positive argument with the amplitude ranging between 0 and 180 is required, the following variation should be used which is based on the formula $R-180\times\times R$ for adding or subtracting 180 depending on whether R is negative or positive.

$$CTOP1 \; : \; \mathbb{Q}(ARG \; R),[.5]((0L,\times 1 \; 0/R)$$
$$\times 180\times\times T)+T\leftarrow(180\div 01)\times\bar{\;}30\div/\phi R$$

Example :

```
     CTOP1 S
1.414    45
2.236  153.4
```

It remains true that for all **X**

$$X \; \leftrightarrow \; PTOC \; CTOP1 \; X \; \leftrightarrow \; CTOP1 \; PTOC \; X$$

6.7. Patterns by Plotting Large Numbers of Points

This is a computer activity which has recently become popular — see for example an article by Barry Martin of Aston University in the Mathematical Gazette, Vol. 70 No. 452 June 1986 pp.140-142 which suggests the iterative computation of equations such as

$$x_{n+1} = y_n - f(x_n)$$
$$y_{n+1} = A - x_n$$

with ongoing plotting of points with these co-ordinates. Here is a program which does this

```
      ∇ L PATTERN R;I;T;X;Y
[1]    X←T←L[1] & Y←L[2] & I←0
[2]    L1:→0 IF L[4]<I←I+1
[3]    X←Y-⍎R
[4]    Y←L[3]-T
[5]    T←X
[6]    (¯4↑L) PIN X,Y
[7]    →L1
      ∇
```

        L : x_0, y_0, A,
            number of iterations,
            elements 5 to 8 = window
        R : equation f(x) in character string form

One suggestion is

```
      0 0 .4 250 ¯1 ¯1 1 1 PATTERN
            '(×X)×(|X)*.5'
```

6. Geometry and Pattern 85

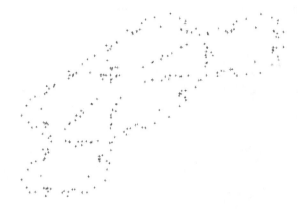

Next experiment by varying L; 0 0 1.9 1000 ⁻4 ⁻4 4 4 produces an interesting "chained links" effect. Then experiment with different functions R. Iterations of as much as 40,000 have been suggested to achieve some patterns.

Chapter 7

Calculus

7.1. Numerical Integration

The following functions AREAS, TRAP, and SIMPSON use the function AXISLAB which is given in Chapter 2, and is now repeated :

$AXISLAB: R[1]+((-/R[2\ 1])\div R[3])\times 0,\iota R[3]$

R : left-hand end of scale,
right-hand end of scale,
number of intervals

7.1.1. Upper and Lower Bounds for Integration

AREAS produces a picture of the upper and lower bounds of integration given by rectangles based on the subdivision of the x range. (see Appendix for BOXES)

```
       ∇ L AREAS R;T;U;X
[1]    X←AXISLAB L
[2]    (L[1],(⌊/T),L[2],⌈/T)BOXES
       (U,1↓T),[1](U←(¯1↓X),0,[1.5]1↓X),¯1↓T↓⍋R
       ∇
```

              L : end-points of x-range,
                      number of points of subdivision
              R : function to be integrated as character string

Example :   0 1 20 *AREAS* '(X*2)×1-X'

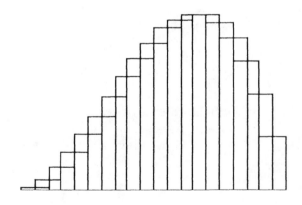

7. Calculus

7.1.2. Trapezium Rule

```
      ∇ Z←L TRAP R;X
[1]   X←AXISLAB L
[2]   Z←((-/L[2 1])÷L[3])
            ×+/((1⌽1 1,(L[3]-1)ρ2)×⍎R)÷2
      ∇
```

        L : range of integration, number of intervals
        R : function to be integrated as character string

First, the values of x are calculated using AXISLAB, then the corresponding y-values are obtained by ⍎R. The estimate of the integral is then the product of half the interval length and the weighted sum of the y-values with weights

    1, 2, 2, ... 2, 1

− the bracketed expression immediately following the +/ is the APL way of describing these weights.

7.1.3. Simpson's Rule

```
      ∇ Z←L SIMPSON R;X
[1]   X←AXISLAB L
[2]   Z←((-/L[2 1])÷L[3])
            ×+/((1⌽1 1,(L[3]-1)ρ4 2)×⍎R)÷3
      ∇
```

        L : range of integration, number of intervals
        R : function to be integrated as character string

This procedure is a more sophisticated version of the Trapezium Rule in which the weights are now

    1, 4, 2, 4, 2, 4, ... 4, 1

and the weighted sum is multiplied by one-third of the interval length. As before the expression in brackets following +/ is an APL description of the weights.

Examples : Integrate x^2 from 0 to 1 using 6 Trapezium/Simpson intervals.

```
      0 1 6 TRAP 'X*2'
0.337963
      0 1 6 SIMPSON 'X*2'
0.333333
```

7.1.4. Adaptive Simpson's Rule

For practical integration, the relevant input parameter to the integrating routine is not how many Trapezium or Simpson intervals should be taken, but rather the degree of precision associated with the answer, i.e. repeated subdivision of intervals should take place until this specified degree of precision is achieved.

Sometimes the integrand may be reasonably "level" in some parts of its range so that a small number of intervals will suffice, whereas it is relatively steep or oscillatory in others where are a much larger number of intervals are required.

A technique to deal with this is the adaptive Simpson rule, where the range of integration is divided into 2 halves, and half of the overall precision is associated with each. The Simpson integrals for each half using 2 and 4 intervals are compared - if they are within the required precision, the latter is accepted as the value, if not, a further subdivision takes place. This is a situation to which recursion naturally applies. Line 2 embodies this stopping test; if the required precision is not achieved, lines 3 and 4 are executed which return the sum of 2 further calls to *ADSIM*.

7. Calculus

```
      ∇ Z←L ADSIM R;T
[1]   Z←(L[1 2],4) SIMPSON R
[2]   →0 IF L[3]>|Z-(L[1 2],2) SIMPSON R
[3]   Z←((L[2]-(T←0.5×-/L[2 1]),0),0.5×L[3])
                                             ADSIM R
[4]   Z←Z+((L[1]+0,T),0.5×L[3]) ADSIM R
      ∇
```

       L : lower limit of integration,
           upper limit of integration,
           absolute precision
       R : function to be integrated as character string

Example : Integrate x^5 from 0 to 1 to an accuracy of 5 decimal places.

```
      0 1 .000001 ADSIM 'X*5'
0.16667
```

7.2. Root Finding

The problem is to find to a specified precision a root of a function of a single variable given in the form $f(x) = 0$.

7.2.1. Bisection Method

In this method the limits of search are halved at every step. It is simple, but slowly convergent.

```
    ∇ L BISECT R;I;T;X;Y
[1]   I←0
[2]   X←L[3 4],.5×+/L[3 4]
[3]   L1:→L3 IF L[1]>|-/2↑Y←⍎R
[4]   →L2 IF 0≠L[2]|I←I+1
[5]   'AT ITERATION ',(⍕I),' X = ',(4↓L)⍕X[1]
[6]   L2:X←X[T],.5×+/X[T←(1+0≠+/×Y[1 3]),3]
[7]   →L1
[8]   L3:'SOLUTION = ',((4↓L)⍕X[1]),
                      ' AT ITERATION ',⍕I
    ∇
```

    L : precision,
        iterations between print,
        lower limit of search,
        upper limit of search,
        two left arguments for ⍕
    R : function in character string form
        for which root is to be found

Example : Find to 4 decimal places a root of $x^2 = 2$ in the range (1,2).

```
    F1←'(X*2)-2'
    .00001 8 1 2 7 4 BISECT F1
AT ITERATION 8 X = 1.4219
AT ITERATION 16 X = 1.4142
SOLUTION = 1.4142 AT ITERATION 19
```

7.2.2. Iteration Method

The principle is that the function whose root(s) are required is expressed in the form $x = g(x)$, and iterations $x_{r+1} = g(x_r)$ are then performed.

7. Calculus

```
      ∇ Z←L ITEROOT R;T;X
[1]   X←L[2]
[2]   →L1 IF L[1]<|L[2]-T←⍎R
[3]   →0,Z←L[2]
[4]   L1:Z←(L[1],⎕←T)ITEROOT R
      ∇
```

               L : precision,
                  start value
               R : function g(x) in character string form
                  for which root is to be found

Note that convergence occurs only when the absolute value of the derivative of the function g is < 1.

Example : Find the roots of $x^2 - 6x + 1 = 0$
2 schemes   (a) $x_{r+1} = (x_r^2 + 1)/6$
                  $g' = x/3$ converges for $-3 < x < 3$

             (b) $x_{r+1} = \sqrt{(6x_r - 1)}$
                 $g' = 3/\sqrt{(6x - 1)}$ converges for $x > 5/3$.

```
      .0001 0 ITEROOT '(1+X*2)÷6'
0.16667
0.1713
0.17156
0.17156
      .005 5 ITEROOT '(¯1+6×X)*.5'
5.3852
5.5956
5.7073
5.7658
5.7961
5.8118
5.8198
5.8198
```

7.2.3. Newton-Raphson Method

This well-known iterative method uses the fact that the tangent to a reasonably smooth curve is an adequate approximation to the curve itself over at least a small region of x.

```
        ∇ L NEWRAP R;I;T;T1;X
[1]     T←(¯1+R⍳';')↑R
[2]     T1←(R⍳';')↓R
[3]     X←L[3]
[4]     I←0
[5]     L1:Z←X-(⍎T)÷⍎T1
[6]     →L3 IF L[1]>|X-Z
[7]     →L2 IF 0≠L[2]|I←I+1
[8]     'AT ITERATION ',(⍕I),' X = ',(3↓L)⍕X
[9]     L2:X←Z
[10]    →L1
[11]    L3:'SOLUTION = ',((3↓L)⍕X),
              ' AT ITERATION ',⍕I
        ∇
```

L : precision,
     iterations between print,
     initial value,
     two left arguments for ⍕
R : function in character string form
     for which root is to be found,
     followed by ';',
     followed by derivative in
       character string form.

Example : Find to 4 decimal places a root of $x^2 = 2$ given $x = 2$ as a first approximation

```
        F2←'(X*2)-2;2×X'
        .00001 2 2 7 4 NEWRAP F2
```

7. Calculus

```
AT ITERATION 2 X = 1.5000
SOLUTION = 1.4142 AT ITERATION 3
```

The following variation estimates the derivative using the first element of L as stepsize and consequently does not require an explicit expression for the derivative:

```
      ∇ L NEWRAP1 R;I;T;X
[1]   X←L[3]
[2]   I←0
[3]   L1:T←⍎R
[4]   X←X+L[1]
[5]   Z←(X-L[1])-T÷((⍎R)-T)÷L[1]
[6]   →L3 IF L[1]>|X-Z
[7]   →L2 IF 0≠L[2]|I←I+1
[8]   'AT ITERATION ',(⍕I),' X = ',
                              (3↓L)⍕X-L[1]
[9]   L2:X←Z
[10]  →L1
[11]  L3:'SOLUTION = ',((3↓L)⍕X-L[1]),
                       ' AT ITERATION ',⍕I
      ∇
```

L : precision,
    iterations between print,
    initial value,
    two left arguments for ⍕
R : function in character string form
    for which root is to be found

Example : Find to 4 decimal places a root of $x^2 = 2$ given $x = 2$ as a first approximation

```
     F2←'(X*2)-2'
```

```
         .00001 2 2 7 4 NEWRAP1 F2
AT ITERATION 2 X = 1.5000
AT ITERATION 4 X = 1.4142
AT ITERATION 6 X = 1.4142
SOLUTION = 1.4142 AT ITERATION 6
```

7.3. Ordinary Differential Equations

Three methods are given, all of which share the same program structure, and differ only in the core statements of the loop. Line 10 is optional and plots the successive points of the iterative sequence.

7.3.1. Euler's Method

```
        ∇ Z←L EUL R;I;N;H;X;Y
[1]     Z←⍳I←0
[2]     X←L[1]
[3]     H←L[3]
[4]     Y←L[4]
[5]     N←⌊(-/L[2 1])÷H
[6]     L1:→0 IF N<I←I+1
[7]     Z←Z,Y
[8]     Y←Y+H×⍎R
[9]     X←X+H
[10]    (0 0,L[2],.4)PIN X,Y
[11]    →L1
        ∇
```

                L : lower limit for x,
                   upper limit for x,
                   step-length H,
                   initial value for y
                R : differential equation in
                   character string form.

7. Calculus

The principle of Euler's method is to propagate values of y using the iterative formula $y_{i+1} = y_i + hf'(x,y)$

The execute function is used in line 8 to evaluate the gradient f' at the left-hand point (x_i, y_i) of the current interval.

Example : Solve the differential equation f' = -xy of the Normal curve, using x = 0, y = $\pi/2$ as start point, and proceeding by steps of .05 to x = 3.

$X \leftarrow .05 \times \iota 60$
$Y \leftarrow (0 \ 3 \ .05, \div(\circ 2)\star .5) EUL \ '-X \times Y'$

7.3.2. Mid-Point Method

Replace lines 8-9 of EUL with

$T \leftarrow Y$
$Y \leftarrow Y + .5 \times H \times \maltese R$
$X \leftarrow X + .5 \times H$
$Y \leftarrow T + H \times \maltese R$
$X \leftarrow X + .5 \times H$

and localize T. Call this function MIDPT.

7.3.3. Trapezium Method

Replace lines 8-9 of EUL with

$T \leftarrow Y$
$Y1 \leftarrow Y$
$Y \leftarrow Y + H \times U \leftarrow \maltese R$
$X \leftarrow X + H$
$Y \leftarrow Y1 + H \times .5 \times U + \maltese R$

and localize T, U and Y1. Call this function **TRAPEZ**.

The aim of **MIDPT** and **TRAPEZ** is to improve on **EUL**. **MIDPT** does so by estimating f′ at the mid-point of the interval (x_i, x_{i+1}), **TRAPEZ** by using the average values of the gradient at the two ends of the interval.

I. Investigate the differences between the three methods by defining

```
X←.05×ι60
Y1←(0 3 .05,÷(○2)*.5)MIDPT '-X×Y'
Y2←(0 3 .05,÷(○2)*.5)TRAPEZ '-X×Y'
```

and then doing the following plots :

```
0 ¯.005 3 .005 PLOT X,[1.5]Y-Y1
0 ¯.005 3 .005 PLOT X,[1.5]Y-Y2
0 ¯.0001 3 .0001 PLOT X,[1.5]Y1-Y2
```

Chapter 8

Probability and Statistics

8.1. Discrete Probability Distributions

8.1.1. Binomial Distribution

The terms in the expansion of

$$(L + (1-L))^R \quad (R \text{ positive integral, } 0 < L < 1)$$

are

$$BINPROB: (L\star T)\times((1-L)\star\phi T)\times(T\leftarrow 0,\iota R)!R$$

Example :

```
      .5 BINPROB 4
0.0625 0.25 0.375 0.25 0.0625
```

HIST (see Appendix 1) demonstrates the above result graphically.

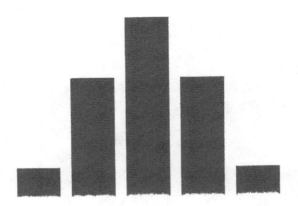

8.1.2. Poisson Distribution

The first L+1 terms for mean R are $e^{-R}R^k/k!$ for $k = 0,..L$, or in APL

```
POISSON:  (*-R)×(R*T)÷!T←0,ιL
```

Comparison with binomial

The difference between a binomial series with n = 10, p = .1, and a Poisson series with mean 1 is shown by

```
      5 POISSON 1
0.3679 0.3679 0.1839 0.06131 0.01533 0.003066

      6↑.1 BINPROB 10
0.3487 0.3874 0.1937 0.0574  0.01116 0.001488
```

8. Probability and Statistics

8.1.3. Hypergeometric Distribution

The hypergeometric distribution describes the probabilities of a sample of size R containing 0,1,2,... defectives, when sampling takes place <u>without replacement</u> from a population of size $L[1]$ of which $L[2]$ are defective.

$HYPGEOM: ((\phi T)!-/L) \times ((T \leftarrow 0, \iota R)!L[2]) \div R!L[1]$

Comparison with binomial

Suppose that we wish to compare the probabilities of drawing 0,1,2,... defectives in a sample of 12 from a box of 40 items of which 10 are defective. The binomial and hypergeometric probabilities are given by

```
      5↑40 10 HYPGEOM 12
0.01548 0.09778 0.242  0.3073 0.22

      5↑.25 BINPROB 12
0.03168 0.1267 0.2323 0.2581 0.1936
```

To make the distinction more explicit, observe that the first terms in the above two series are

$(30/40) \cdot (29/39) \cdot (28/38) \ldots \quad (19/29)$

and $(30/40)^{12}$ respectively, and the third terms are

$$\frac{(_{30}C_{10}) \cdot (_{10}C_2)}{_{40}C_{12}}$$

and $(_{12}C_2) \cdot (0.75)^{10} \cdot (0.25)^2$ respectively. ($_nC_r$ is the number of combinations of r objects chosen from n.)

8.2. The Birthday Problem

The general problem is to calculate the probabilities of finding a pair of people with coincident birthdays in groups of 1,2,3,... people.

We generalise this problem by defining a program which returns the probabilities of finding a matching pair of items in groups of 1,2,3,... L given that each item possesses independently one of a set of values from 1 to R.

      $MATCHPR: \ 1-\times\backslash(R-0,\iota L-1)\div R$

Example :

      ¯2↑23 $MATCHPR$ 365
0.4757 0.5073

i.e. 23 is the size of the smallest group of people for whom there is a greater than 50% chance of a birthday coincidence.

Often the object of the birthday problem is to discover how big the group must be before the probability of finding a matching pair exceeds one half. Use

      30 $MATCHPR$ 365

Then use the following function to simulate the number of matching pairs in a group of L people (R as for **MATCHPR**).

      $MATCHES: \ +/\sim(R\iota R)=\iota\rho R\leftarrow ?L\rho R$

Example :

      (23 $MATCHES$ 365),23 $MATCHES$ 365
2 1

8. Probability and Statistics

8.3. Descriptive Statistics

The standard idiom for average of a vector as given in Appendix 2 is $(+/R)\div\rho R$. Here is a version which also produces column averages if R is a matrix :

$$AV : (+\not/R)\div 1\uparrow\rho R$$

For data in vector or matrix form, the following phrase subtracts the mean from all values of vector R, or column means from all columns when R is a matrix:

$$MADJ : R-(\rho R)\rho AV\ R$$

This is then used to calculate variance and standard deviation.

Examples :

```
      ⌽X
3   2   2   4   6   7   7   9
25 14 18 27 35 40 32 53

      AV X
5 30.5
      ⌽MADJ X
 ¯2     ¯3    ¯3    ¯1    1    2    2    4
¯5.5 ¯16.5 ¯12.5 ¯3.5  4.5  9.5  1.5 22.5
```

8.3.1. Variance

$$VAR : (+\not/(MADJ\ R)\star 2)\div 1\uparrow\rho R$$

In the case of sampling variance, i.e. where the divisor is (n-1) rather than n, use

$$SVAR : (+\not/(MADJ\ R)\star 2)\div {}^{-}1+1\uparrow\rho R$$

8.3.2. Standard deviation

           SD : (VAR R)*.5

Examples :

           ⌽X
     3 2 2 4 6 7 7 9
    25 14 18 27 35 40 32 53

           VAR X
    6 136.3
           SVAR X
    6.857 155.7
           SD X
    2.449 11.67

8.3.3. Partition Values

Median and Percentiles for vector R

The algorithm for MEDIAN must take account of the fact that if the number of items is odd, the value will be that of the single middle item following sorting, whereas if the number of items is even, the median value is half the sum of the two middle items.

           MEDIAN : .5×+/R[⌈.5×0 1+⍴R←R[⍋R]]

Following the function SEP in Chapter 2, the percentile algorithm uses encode (⊤) in line 1 to separate a mixed number representing the position value of the required percentile into its integer and fractional parts. The fractional part gives how far between two items in the ordered data the required item lies, and the resulting calculation of the percentile value itself takes place in line 2.

8. Probability and Statistics

```
        ∇Z←L PCTILE R
[1]     Z←,0 1⊤1+.01×L×⁻1+ρR←R[⍋R]
[2]     Z←R[Z[1]]+Z[2]×-/R[Z[1]+1 0]
        ∇
```

        L : percentile (integer)
        R : data (vector)

Examples :

```
      ⌽X
3  2  2  4  6  7  7  9
25 14 18 27 35 40 32 53

      MEDIAN X[;1]
5
      25 PCTILE X[;2]
23.25
```

8.3.4. Mode and Range

Range of a vector, or column ranges of a data matrix

     RANGE : (⌈⌿R)-⌊⌿R

Mode of a vector R

The mode is a vector (possibly with just one item) of the items in R which have the highest frequency. An outer product is used to match the data items against the set of unique items. Adding columns counts the frequencies of the unique items, and a reduction following comparison with the largest such sum yields the (not necessarily unique) mode. REMDUP is in Appendix 2.

     MODE : ((⌈/T)=T←+⌿R∘.=Z)/Z←REMDUP R

Examples :

```
      ⌽X
3  2  2  4  6  7  7  9
25 14 18 27 35 40 32 53

      RANGE X
7 39
      MODE X[;1]
2 7
```

8.4. Random numbers from Various Distributions

R random numbers from uniform distribution in the interval (0,1):

$$RND : (?T) \div T \leftarrow R\rho\lfloor/\iota 0$$

L random numbers from uniform distribution in the interval (R[1],R[2]):

$$RUN : R[1]+(RND\ L)\times-/R[2\ 1]$$

L random numbers from negative exponential distribution mean R:

$$RNE : -R\times\circledast RND\ L$$

L random numbers from Normal distribution, mean R[1] and standard deviation R[2], using the Box-Muller formula:

$$RNO : R[1]+R[2]\times((2\times L\ RNE\ 1)\star.5)\times 2 \circ o2\times RND\ L$$

L random Booleans (i.e. 0 or 1) where constant probability of drawing a 1 is R:

$$RBO : R>RND\ L$$

8. Probability and Statistics

L random variables from the set $\iota\rho R$ whose respective probabilities are R, and also $+/R \leftrightarrow 1$:

$$RSA : +/(RND\ L)\circ.\geq 0,\bar{}1\downarrow+\backslash R$$

Frequency distribution for a sample such as the above:

$$RFD : +/(\iota\rho R)\circ.=\lceil RSA\ R$$

Examples :

```
      RND 5
0.5133 0.846 0.8415 0.4154 0.4679

      4 RUN 10 20
11.78 15.72 10.33 14.98

      4 RNE 10
2.9 1.157 1.719 15.48

      4 RNO 10 3
14.65 12.96 10.74 8.39

      10 RBO .75
1 1 0 0 1 1 1 1 0 1

      10 RSA .4 .3 .1 .2
1 4 2 1 3 1 2 2 4 3

      100 RFD .4 .3 .1 .2
35 33 6 26
```

8.5. Simulations

8.5.1. Dice/Coins,etc.

The result of R throws of a single die is given by

   $DICE\ :\ ?R\rho 6$

Example :

   $DICE\ 4$
$4\ \ 1\ \ 2\ \ 4$

In a similar way $?R\rho 2$ simulates R throws of a single coin.

Multiple dice, i.e results of L sums of throws of R dice

   $MDICE\ \ :\ \ +/?(L,R)\rho 6$

Example :

   $6\ MDICE\ 2$
$8\ \ 6\ \ 10\ \ 12\ \ 5\ \ 6$

8.5.2. Buffon's needle

   $BUFFON\ :\ +/(RND\ R)<2oR\ RUN\ oO\ .5$

This program simulates R throws of a needle of length 2 on a striped flag with stripes of length 2. The result is the number of throws in which a needle cuts a stripe. The distance x of the centre of the needle from the nearest stripe boundary is therefore uniformly distributed between 0 and 1. The angle y between the needle and the direction of the stripes varies uniformly between 0 and 90 degrees. The needle cuts a stripe if x < sin y. For a fixed x this happens $\sin^{-1}x/\pi/2$ of the time and hence as x ranges from 0 to 1, the proportion of throws which cut a stripe is on average the integral of sin⁻¹x/ $\pi/2$ from 0 to 1 which = $2/\pi$ = .637.

Examples :

```
      (BUFFON 100),(BUFFON 100),BUFFON 100
58 64 60
```

8.6. Frequency Distributions

```
      FREQ  :  +/(⌈(,R-L[1])÷L[2])∘.=⍳L[3]
```

        L : lower boundary of leftmost cell,
           cell-width,
           number of cells
        R : numeric vector

Example :

```
      0 2 5 FREQ 3 2 2 4 6 7 7 9
2 2 1 2 1
```

Note that with the above form of FREQ, values on a cell boundary go into the cell on the left. If the opposite convention is required, replace ⌈ with MAXI :

```
      MAXI  :  (R=⌈R)+⌈R
```

Often, when data is given in the form of a frequency table, it is useful to transform it into a single vector with repeated values. Do this using replicate with the frequencies as left argument, e.g.

```
      3 2 2/1 2 3
1 1 1 2 2 3 3
```

If replicate is not implemented, use the function REPLIC in Appendix 2.

8.6.1. Stem-and-leaf Plot

Another way of summarising data in a semi-pictorial way is the so-called "stem-and-leaf" plot. The following program produces a simple form of this plot for a vector R of positive numbers. An estimate V is made in line 2 of a suitable number of stems.

```
        ∇ STMLEAF R;I;T;U;V;W
[1]     T←-⌊10⍟(RANGE R)÷10×10⍟⍴R
[2]     V←.1×⌈/R←(T ROUND R)×10*T
[3]     U←1+(0⌈-T)+⌈10⍟V
[4]     I←¯1+⌊/⌊.1×R
[5]   L1:→0 IF V<I←I+1
[6]     ((((U+1),0)+0⌈T)⍕I×10*-T-1),'|',
                1 0⍕W[⍋W←10|(I=⌊.1×R)/R]
[7]     →L1
        ∇
```

Example :

```
      STMLEAF 4.9 2.7 3.6 5.1 2.8 2.8 3.0
                     4.6 5.2 5.8 2.2
2.0|2788
3.0|06
4.0|69
5.0|128
```

8.6.2. Two-way Frequency Distribution

When data is classified according to two variables, the function TWOWAY produces a 2-dimensional frequency table.

```
       ∇ Z←L TWOWAY R;I
[1]    I←0  &   Z←(L[1;3],0)ρ0
[2]    L1:→0 IF L[2;3]<I←I+1
[3]    Z←Z,L[1;] FREQ(R[;2]INRANGE
              L[2;1]+L[2;2]×⁻1 0+I)/R[;1]
[4]    →L1
       ∇
```

    L : matrix with two rows, each of which has
      3 items defined as per L of FREQ
    R : matrix with 2 columns - each row corresponds
      to an observation, and each column to one
      of the two classifications defined by
      the rows of L.

Example :

```
       ⍉X
 3  2  2  4  6  7  7  9
25 14 18 27 35 40 32 53
       L
 0  2  5
10 10  6

       L TWOWAY X
2 0 0 0 0 0
0 2 0 0 0 0
0 0 1 0 0 0
0 0 2 0 0 0
0 0 0 0 1 0
```

TWOWAY uses an auxiliary function INRANGE which for a numeric vector L returns a binary vector of the same length with a 1 if the corresponding element of L is within the range defined by R.

    INRANGE : $(+/\times L \circ . - R) \in 0\ 1$

Example :

    $(\iota 10) INRANGE\ 5\ 7$
0 0 0 0 0 1 1 0 0 0

If MAXI was used in FREQ change 0 1 to 0 ¯1. A simple extension of this function returns those elements of L which are within the range R :

    WITHIN : $(L\ INRANGE\ R)/L$

Example :

    $(\iota 10) WITHIN\ 5\ 7$
6 7

Change the 0 1 of INRANGE to 0 for strict inclusion, or to ¯1 0 1 if both end-points are to be included.

8.6.3. Scatterplots

Scatterplots in general are given by

    window *PLOT* $Y, [1.5] X$

where Y and X are vectors of paired values.

8. Probability and Statistics

Pictorial impressions of sampling from the various distributions can be obtained by using PLOT or PIN. You will find yourself forced to use the latter if the number of points plotted is large enough to force WS FULL.

To display a one-dimensional distribution, impose a narrow Normal distribution on the y-axis so that there is some separation of the points corresponding to the distribution of interest on the x axis, e.g. to display a sample of 100 random Normal values from N(10,3) use

```
1 ¯1 19  1 PLOT (100 RNO 10 3),[1.5]
                       100 RNO 0 .02
```

Two-dimensional distributions can be shown using exactly the same phrase, e.g. changing the window to 1 ¯.06 19 .06 gives a display of a bivariate Normal distribution. Substituting DRAW for PLOT gives zig-zag patterns which arise from combining different types of joint distributions (Normal vs. exponential, Normal vs. uniform, etc.)

8.7. Regression

Fitting polynomials of varying degree to a set of data

$$POLYFIT : L[;2] \boxplus \phi L[;1] \circ . \star 0, \iota R$$

L : 1st. column = x-values,
    2nd col   = y-values
R : degree of polynomial

Estimates of y based on above polynomial regression

$$EST : L[;1] \ POLY \ L \ POLYFIT \ R$$

(see Chapter 3 for POLY)

Residuals following above polynomial regression

   RESID : L[;2]-L EST R

Example :

   ⍒*X*
```
 3  2  2  4  6  7  7  9
25 14 18 27 35 40 32 53
```

   X POLYFIT 1
4.452 7.792

i.e. best fitting straight line is y = 4.452x + 7.792.

   X EST 1
```
21.42 16.88 16.88 25.96 35.04 39.58
                              39.58 48.67
```

   X RESID 1
```
3.583 ¯2.875 1.125 1.042 ¯0.04167 0.4167
                              ¯7.583 4.333
```

The following function plots the least squares fit (i.e. the original points, the regression line and the residuals) :

```
      ∇ Z←L PDRAW R;I
[1]   L DRAW (ι1↑ρR),[1.5]R EST 1
[2]   I←0
[3]   L1:→0 IF (1↑ρR)<I←I+1
[4]   L PIN R[I;]
[5]   L JOIN R[I;1],(R EST 1)[I]
[6]   →L1
      ∇
```

     L : window
     R : data matrix, as per L of POLYFIT

8. Probability and Statistics

8.8. Correlation

Product moment correlation coefficient

This is sometimes known as the Pearson coefficient and is given algebraically by

$$\frac{\Sigma (x - \mu_x)(y - \mu_y)}{\sqrt{\Sigma(x-\mu_x)^2} \cdot \sqrt{\Sigma(y-\mu_y)^2}}$$

where μ_x and μ_y are the means of x and y respectively.

In APL

$$CCOF : (+/(MADJ\ R) \times MADJ\ L) \div$$
$$\times/((SS\ R), SS\ L) \star .5$$

   L : numeric vector
   R : numeric vector of equal length

where the auxiliary function SS returns the sum of squares of a vector R about its mean

$$SS : +/(MADJ\ R) \star 2$$

8.8.1. Covariance and Correlation matrices

$$COVM : (\lozenge R) + . \times MADJ\ R$$

$$CORM : (COVM\ R) \div (\lozenge T) + . \times T \leftarrow (1, {}^-1 \uparrow \rho R) \rho$$
$$(+ \neq (MADJ\ R) \star 2) \star .5$$

   R : (NxP) data matrix with rows corresponding to observations and columns to variables

The items on the leading diagonal of the covariance matrix are the sums of squares about the mean for each of the variables. The remaining items are the various cross-products of the form
$\Sigma\ (x - \mu_x)(y - \mu_y)$

The correlation matrix gives the product moment correlation coefficients between all pairs of variables in the form of a square symmetric matrix.

Example :

```
      ⌽X
 8  7  6  5  4  3  2  1
 3  2  2  4  6  7  7  9
25 14 18 27 35 40 32 53

      X[;2] CCOF X[;3]
0.9531

      COVM X
  42  ¯42 ¯180
 ¯42   48  218
¯180  218 1090

      CORM X
 1       ¯0.9354 ¯0.8413
¯0.9354   1       0.9351
¯0.8413   0.9351  1
```

8. Probability and Statistics

8.9. Non-parametric Tests

8.9.1. Runs Test

The following function counts the number of runs in a sequence

$$RUNS\ :\ (R[1]=R[\rho R])++/R\neq 1\phi R$$

$$R\ :\ \text{numeric vector}$$

Example :

```
    RUNS 'HHHHTHTTTHHTTTTHHHHH'
7
```

If there are only two types of item in the sequence, the value of RUNS can be used to determine whether the sequence is consistent with a null hypothesis that each item is a random drawing of one of the two types.

The following is a table of critical values :

```
 ρR =   10  20  30  40  60  80 100 120 140 160 180 200
 sig   ------------------------------------------------
 2.5%    2   6  10  14  22  31  40  49  58  68  77  86
   5%    3   6  11  15  24  33  42  51  60  70  79  88
  95%    8  15  20  26  37  48  59  70  81  91 102 113
97.5%    9  15  21  27  39  50  61  72  83  93 104 115
```

The 7 runs observed in the sequence of 20 characters above are not inconsistent with a hypothesis of randomness.

8.9.2. Rank Correlation

There are two well known coefficients which test how well two rankings match. (The term "ranking" is used to denote "rank ordering" to avoid confusion with rank in the APL sense of $\rho\rho R$.)

To be specific, let us suppose that the marks in order of 10 students in French and German are given by

$FREN \leftarrow$ 75 28 32 47 83 92 69 19 45 78
$GERM \leftarrow$ 74 52 41 72 80 86 73 35 67 89

so that the rankings are:

⍋⍋$FREN$
7 2 3 5 9 10 6 1 4 8
⍋⍋$GERM$
7 3 2 5 8 9 6 1 4 10

8.9.2.1. Spearman's coefficient

If d_i is the difference in rankings for the ith. student and N is the total number of students, Spearman's coefficient is defined as

$$1 - 6\Sigma d_i^2 / N(N^2 - 1)$$

In APL, if L and R are two sets of rankings, this coefficient is

$SPEAR$: $1-(6\times+/(L-R)\star 2)\div(T\star 3)-T\leftarrow\rho R$

(⍋⍋$FREN$)$SPEAR$ ⍋⍋$GERM$
0.9515

8. Probability and Statistics

The following is an abridged table of critical values for a two sided-test :

N =	5	6	7	8	9	10	12	14	16	18	20
sig											
5%	1.00	.89	.79	.74	.70	.65	.59	.54	.50	.47	.45
10%		.90	.83	.71	.64	.60	.56	.50	.46	.43	.40	.38

Thus for $N=10$, the probability of the Spearman coefficient exceeding .65 is $½ \times 5 = 2½\%$. The conclusion for the marks data therefore is that student performances the two subjects are highly correlated.

8.9.2.2. Kendall's Coefficient

Continuing the example of the previous section, every one of the $_{10}C_2$ pairs of students can be assessed as either having consistent rankings between subjects (scoring $+1$), or as having inconsistent rankings (scoring -1). Here again are the rankings:

```
    ♣♣FREN
7 2 3 5 9 10 6 1 4 8
    ♣♣GERM
7 3 2 5 8 9 6 1 4 10
```

Students 1 and 2 are consistent in that student 2 excels student 1 in both subjects. Students 2 and 3 however are inconsistent since student 2 excels student 3 in French but the reverse is true for German. The Kendall coefficient is

$$\frac{\text{(total plus scores)} - \text{(total minus scores)}}{_nC_2}$$

In APL, if L and R are two sets of rankings, this coefficient is

$$KENDALL: \; +/,((UPT \; \rho R) \times (\times L \circ .-L) \times \times R \circ .-R) \\ \div 2! \rho R$$

(see Appendix 2 for UPT)

Example :

        (⍋⍋FREN)KENDALL ⍋⍋GERM
0.8667

The following is a table of critical values for a two sided-test :

```
    N =   5    6    7    8    9   10   12   14   16   18   20
sig      ------------------------------------------------------
 5%     1.00  .87  .71  .64  .56  .51  .45  .41  .38  .35  .33
10%      .80  .73  .62  .57  .50  .47  .39  .36  .32  .29  .27
```

Like the Spearman test, this test shows high correlation between the performances of the students in French and German.

More Non-parametric Statistics

The next three tests have the general aim of testing samples as a whole for consistency, i.e. the null hypothesis is that the samples have been drawn from the same population. For the marks example, they derive measures for the consistency of marking levels across <u>subjects</u> rather than correlation between subject marks of individual <u>students.</u> In each case L and R are the two vectors to be compared by the test, i.e. the marks themselves, as opposed to their rankings.

8.9.3. Sign Test

STEST returns a 2-item vector, the first item being the number of cases in which the L value exceeds the R value, and the second the number of cases in which it is less than the R value. It is assumed therefore that the items in L and R are paired, and thus that L and R are of equal length.

        STEST : (+/0⌈T),+/-0⌊T←×L-R

8. Probability and Statistics

The null hypothesis is that there is no difference in the underlying populations from which L and R are drawn, and hence the probabilities for each I that $L[I]>R[I]$ and $L[I]<R[I]$ are both one half.

Thus to assess the result of the test compute

            SPROB : +/(1+L/L STEST R)↑.5 BINPROB ⍴L

which is the probability of obtaining the observed value or less at even luck. If this is sufficiently low, L and R may be judged significantly different. Alternatively use the following table of 2-tailed critical values for L/L STEST R. The following table gives the maximum values for which the result of STEST is significant at the level quoted, e.g. for N = 10, the probability of the result being 1 or 0 is less than .025.

```
  N =    7   8   9  10  12  14  16  18  20  22  24
sig     ---------------------------------------------
2.5%    0   0   1   1   2   2   3   4   5   5   6
5%      0   1   1   1   2   3   4   5   5   6   7
```

Example :

```
         FREN STEST GERM
3 7
         FREN SPROB GERM
0.1719
```

The conclusion is thus that there is no significant difference between mark levels in French and German.

8.9.4. Wilcoxon Signed Rank Test (W-test)

For this test the non-zero absolute differences (L-R) are ranked (see line 1). Each such ranking is then given a positive sign if L is greater than R, and a negative sign if L is less than R. The W-statistic is the smaller of the sums of the positive and negative rankings and its (1-tailed) significance can be assessed from the table which follows.

```
        ∇ Z←L WTEST R;T
[1]     T←(×T)×TRANKU |T←(L-R)WITHOUT 0
[2]     Z←L/(+/0⌈T),+/-0⌊T
        ∇
```

(see Appendix 2 for WITHOUT and TRANKU)

ρL =	7	8	9	10	12	14	16	18	20	22	24
sig											
0.5%	0	0	1	3	7	12	19	27	37	48	61
2.5%	2	3	5	8	13	21	29	40	52	65	81

Differences are significant if the result of the test is <u>below</u> the values given above. With the marks data the steps in the calculation are as follows:

```
        FREN
75 28 32 47 83 92 69 19 45 78
        GERM
74 52 41 72 80 86 73 35 67 89
        |FREN-GERM
1 24 9 25 3 6 4 16 22 11
        ▲▲|FREN-GERM
1 9 5 10 2 4 3 7 8 6
```

The 3 cases in which a French mark exceeds the corresponding German mark correspond to the entries 1, 2 and 4 in the last vector, which total 7. Hence:

```
        FREN WTEST GERM
7
```

8. Probability and Statistics

This is significant at the 2½% level, i.e. the probability of obtaining a value of 7 or lower is less than .025. The W-test is inherently more sensitive than the sign test since it uses <u>values</u> rather than just signs. In the present example, it leads to the conclusion that there <u>is</u> a significant difference between the mark levels of French and German, as opposed to the Sign Test which did not.

8.9.5. Mann-Whitney Rank Sum Test (U-test)

This test is a non-parametric equivalent of the t-test applied to a pair of <u>independent</u> (and thus unpaired) samples. L and R are ranked collectively (see line 1), and the U-statistic is based on the smaller of (1) the total rankings corresponding to items in L, and (2) the total rankings corresponding to items in R.

```
       ∇ Z←L UTEST R;T
[1]    T←TRANKU L,R
[2]    Z←L/((+/(ρL)↑T),+/(ρL)↓T)-2!1+(ρL),ρR
       ∇
```

For the student marks

```
       ⍋⍋FREN,GERM
14 2 3 7 17 20 10 1 6 15 13 8 5 11 16 18
                                  12 4 9 19
```

The sum of the first 10 rankings corresponding to French is 95, and of the second 10, corresponding to German, 115. The U-statistic is obtained by subtracting the minimum possible rank sum from each. Since the numbers of French and German students are both 10 each minimum is 1 + 2 + 3 + ... + 10 = 55.

L and R need not be of equal length for this test. When they are, and $N = \rho L$, significance levels are given by the following table which shows for example that for $N=10$, the probability of U being 16 or less is less than .005.

```
  N =    5    6    7    8    9   10   12   14   16   18   20
  sig  ---------------------------------------------------------
  0.5%   0    2    4    7   11   16   27   42   60   81  105
  2.5%   2    5    8   13   17   23   37   55   75   99  127
```

If L and R are of unequal length, consult more detailed statistical tables. Running *UTEST* with the marks data we have

```
        FREN UTEST GERM
40
```

This is greater than 23 and so there is no reason to reject the null hypothesis that similar marking standards apply to the two subjects. Comparing this with the W-test shows that as <u>paired</u> data, the marks show a significant difference between subjects. On the other hand, if the marks are regarded as 2 sets of independent samples no such evidence is present.

If all the French marks are increased by 25, and all the German marks lowered by 25 the U-test gives

```
        (FREN+25) UTEST GERM-25
11
```

The internal rankings and thus the Spearman and Kendall correlation coefficients are unchanged, but UTEST now indicates significant discrepancy between the mark levels in the two subjects.

8. Probability and Statistics

8.9.6. Goodness of Fit

The formula for comparing a vector L representing observations with an equal length vector R of expected values is

$$\Sigma\ (O-E)^2/E$$

or in APL

$$GOFIT\ :\ +/((L-R)\star 2)\div R$$

Significance testing is then carried out using chi-squared tables (see below). To apply Yates' correction only a small adjustment to GOFIT is necessary.

$$YATES\ :\ +/((.5-|L-R)\star 2)\div R$$

To calculate a table of expected values for given marginal distribution vectors L and R as in the example given below use

$$EXPEC\ :\ (L\circ.\times R)\div +/R$$

Example :

```
      6 1⌽107 43 EXPEC 61 51 38
43.5   36.4   27.1
17.5   14.6   10.9
```

In using this function it is prudent to include a test that $(+/L)\leftrightarrow(+/R)$.

Given a two-way frequency table R, one can go straight to the chi-squared value by

$$CHISQ\ :\ (,R)GOFIT,(+/R)EXPEC\ +/[1]R$$

Example : The following results emerged from an experiment in which the effectiveness of a treatment was to be judged by counting the number of good, bad and unaffected items in two sets of experimental units, one of which received the treatment and the other not :

```
              Good  Unaffected  Bad  Totals
    Treated    46       32       29    107
Not treated    15       19        9     43
           ---------------------------------
               61       51       38    150
```

The appropriate value of chi-squared is obtained by

```
       CHISQ 2 3ρ46 32 29 15 19 9
2.797
```

8.10. Statistical Tables

Normal, t, chi-squared, and F tables can be obtained using the equations of the appropriate probability density functions and integrating these using the *ADSIM* function given in Chapter 7. It is useful first to establish a routine for the Beta function with parameters given by the 2-item vector R :

```
       BETA : (×/!R−1)÷!+/¯1,R
```

Note that $\Gamma(\frac{1}{2}) = \sqrt{\pi}$, so $B(\frac{1}{2},\frac{1}{2}) = \pi$.
Also $\Gamma(1\frac{1}{2}) = \frac{1}{2}\sqrt{\pi}$, $B(1\frac{1}{2},\frac{1}{2}) = \frac{1}{2}\pi$,
$\Gamma(2\frac{1}{2}) = 1\frac{1}{2}.\frac{1}{2}\sqrt{\pi}$, $B(2\frac{1}{2},\frac{1}{2}) = \frac{3}{8}\pi$, etc.

8.10.1. Normal probability density function

Function is $(1/\sqrt{(2\pi)})\exp(-\frac{1}{2}x^2)$

```
       NPDF : (÷(○2)*.5)××¯.5×R*2
```

R : x-value or vector of x-values

8. Probability and Statistics

Example : What is the probability of a random Normal variable lying between 0 and 2 standard deviations from the mean? One way to find this is to use Simpson's formula to integrate the Normal p.d.f.:

```
      0 2 .0001 ADSIM 'NPDF X'
0.4772
```

Another way to calculate Normal probabilities is to use a closely approximating formula for the Normal integral given in "Calculation of the Normal distribution Function," P.A.P. Moran, Biometrika, 1980, 67.3, pp.675-6:

```
      NINT:.5+((R÷3×2*.5)++/(((ρR)ρ0)∘.+
      (*-T×T÷9)÷T)×1O(R∘.×(2*.5)×T←⍳12)÷3)÷O1

      NINT ¯1 1 2
0.1587 0.8413 0.9772
```

The inverse problem of finding the Normal Rth. percentile (or vector of percentiles) is solved to a high degree of approximation by an algorithm in "Approximations for Digital Computers," C. Hastings, Princeton University Press, 1955:

```
      NPCT : T-((T∘.*0 1 2)+.×C)÷
             1+((T←(⍟÷(1-.01×R)*2)*.5)∘.*⍳3)+.×D
```

where $C \leftrightarrow 2.515517\ 0.802853\ 0.010328$
           $D \leftrightarrow 1.432788\ 0.189269\ 0.001308$

Example :

```
      NPCT 95 97.5 99.5
1.645 1.96 2.807
```

8.10.2. Student t probability density function

Function is $1/B(\frac{1}{2}, \frac{1}{2}n) (\sqrt{n})(1 + (x^2/n))^{(n+1)/2}$

t_1 is $1/\pi(1 + x^2)$
t_2 is $1/2\sqrt{2}(1 + \frac{1}{2}x^2)^{1.5}$
t_3 is $2/\pi\sqrt{3}(1 + x^2/3)^2$, etc.

```
TPDF:÷(L*.5)×(BETA T)×
   (1+(R*2)÷L)*+/T←.5×1,L
```

L : degrees of freedom
R : x-value or vector of x-values

Example : What is the probability of a random t_2 variable lying between 0 and 2 standard deviations from the mean?

```
      0 2 .0001 ADSIM '2 TPDF X'
0.4082
```

The inverse problem of finding the percentiles of t for L degrees of freedom is approximated by an algorithm due to K.J. Koehler - see Technometrics, Vol. 25 No.1, 1983. Either L or R, but not both, may be vectors.

```
TPCT:÷¯.0953+(¯.631÷1+L)+
   (.81×(-⊛R×2-R)*¯.5)+.076×
((R←2×1-.01×R)×(¯.5+(○2)*.5)×L*.5)*÷L
```

Examples:

```
      1 10 100 TPCT 97.5
7.83 2.242 1.979
      2 TPCT 95 97.5 99.5
2.752 3.939 8.588
```

8. Probability and Statistics

8.10.3. Gamma and Chi-squared probability density functions

The Gamma p.d.f. is $(x^{a-1}e^{-bx}b^a)/\Gamma(a)$.

$$GPDF : (R\star U)\times(\star-T\times R)\div(!U\leftarrow 1\uparrow L-1)\div$$
$$(T\leftarrow 1\downarrow L)\star 1\uparrow L$$

> L : shape & scale parameters (a,b) respectively
> R : x-value or vector of x-values

X^2 is a special case of Gamma with scale parameter ½. The parameter, degrees of freedom, is twice the corresponding Gamma shape parameter.

$$CPDF : (.5\times L,1)GPDF\ R$$

> L : degrees of freedom
> R : x-value or vector of x-values

The first few X^2 curves are:
$X^2_1 : (1/\sqrt{(2\pi)})x^{-1/2}e^{-x/2}$
$X^2_2 : \frac{1}{2}e^{-x/2}$
$X^2_3 : (1/\sqrt{(2\pi)})x^{1/2}e^{-x/2}$
$X^2_4 : \frac{1}{4}xe^{-x/2}$

Example : What is the probability of a random chi-squared variable lying between 0 and 11 for 5 degrees of freedom?

```
      0 11 .001 ADSIM '5 CPDF X'
0.949
```

Chi-squared percentiles can be obtained by first converting the corresponding Normal percentile using a formula called the Wilson-Hilferty approximation:

$$CPCT:L\times(1-(2\times\div 9\times L)-$$
$$((NPCT\ R)\div 3)\times(2\div L)\star .5)\star 3$$

L : degrees of freedom (scalar or vector)
R : required percentile (or vector of %tiles)

Examples:

```
      1 2 CPCT 95
3.748 5.938
      1 CPCT 90 95
2.639 3.748
```

Note that accuracy falls off rapidly when R is less than about 20.

8.10.4. F probability density function

Function is $\dfrac{a^{a/2} b^{b/2} x^{(a-2)/2}}{B(\frac{1}{2}a, \frac{1}{2}b)(ax+b)^{\,x^{(a+b)/2}}}$

```
      ∇ Z←L FPDF R
[1]   Z←(×/L*.5×L)×R*⁻1+.5×1↑L
[2]   Z←Z÷(BETA .5×L)×(L[2]+L[1]×R)*.5×+/L
      ∇
```

L : degrees of freedom (2-item vector)
R : x-value or vector of x-values

Some F curves are:

$F_{(1,1)}$: $1/\pi(\sqrt{x})(x+1)$ $F_{(2,1)}$: $(2x+1)^{-1.5}$
$F_{(2,4)}$: $8(x+2)^{-3}$ $F_{(4,1)}$: $12x(4x+1)^{-2.5}$

Example : What is the probability of a random F-variable lying between 0 and 5 for $F_{(5,6)}$?

```
      0 5 .001 ADSIM '5 6 FPDF X'
0.962
```

8.11. Sample Sizes

Consider the problem of calculating the sample size necessary to estimate the proportion of a population possessing a given attribute subject to a predeclared confidence level that the true but unknown value should be within a fraction S on either side of the estimate. The formula is

$$\frac{P(1-P)}{(S/Z)^2}$$

where P is an "a priori" estimate of the population proportion (the worst case arises, other factors being equal, when P is 0.5 — the closer P moves to 0 or 1, the smaller the required sample size), and Z is the standard Normal percentile corresponding to the required confidence interval in a double sided sense. For example 95% and 99% confidence levels have Z-values of 1.96 and 2.58 respectively.

For given ranges of P, S and Z, sample size tables are obtained by:

```
      P←.5 .1
      S←.01 .02 .05
      Z←1.96 2.58

      +SSIZE←⌈(P×1-P)∘.÷(S∘.÷Z)*2
9604 16641
2401  4161
 385   666

3458  5991
 865  1498
 139   240
```

Planes correspond to values of P, columns to confidence levels, so if the anticipated P could be as much as 0.5, and the sample estimate is required to be within .02 of the population value on either side, then sample sizes of 2401 and 4161 are required for 95% and 99% confidence levels respectively.

If the population size N is <u>not</u> indefinitely large, the sample size reduces according to the formula

$$\frac{P(1-P)}{(S/Z)^2 + P(1-P)/N}$$

The convenient way to handle this is to reciprocate the previous table, adjust, and reciprocate:

```
      ⌈÷(÷SSIZE)∘.+÷N←5000 10000
3289 4900
3845 6247

1623 1937
2272 2939

 358  371
 588  625


2045 2570
2726 3747

 738  797
1153 1303

 136  138
 230  235
```

8. Probability and Statistics 133

An extra dimension corresponding to N has been introduced. It features as columns in the above display so that for example 2401 should be adjusted to 1623 and 1937 for populations of 5000 and 10000 respectively, and 4161 should be adjusted to 2272 and 2939. In interpreting a table of this sort the user must assess carefully the validity of the assumptions which underlie the basic formula, particularly those concerning random sampling.

Chapter 9

Combinatorics

Combinatorial algorithms deal with the arrangement of objects in cells. The number of such algorithms is legion due to such considerations as : are the objects ordered within cells; are the cells themselves ordered; are empty cells allowed; how many kinds of distinguishable objects are present. The whole range of such algorithms would fill a book in its own right − a few of the most basic ones are given here covering permutations (with derangements as a subset), combinations, selections with repetition, compositions and partitions.

9.1. Permutations in Lexical Order

The following algorithm gives a list of all permutations of first R integers in lexical order.

The first step is to define a function which transforms an integer R into its representation in a number base whose elements are $!\phi\iota L$:

```
FACTINT : (φ1+ιL)⊤R
```

Example :

  3 `FACTINT` 21
3 1 1

i.e. 21 is represented in 3-factorial notation by the vector 3 1 1 since 21 = (3×3!) + (1×2!) + (1×1!). Provided that the successive elements of such a vector do not exceed $\phi \iota M$, then each of the integers $^{-}1+\iota\ !M+1$ is uniquely represented.

Suppose that we want to obtain the Ith. permutation of ιN when the permutations are arranged in lexical ordering, i.e. the ordering which for N = 4 begins

 1 2 3 4 1 2 4 3 1 3 2 4 ....

A possible procedure is to start with ιN alongside the (N-1)-factorial representation of I, the successive digits of which are then used as indexes <u>in origin 0</u> to pick out the elements of the permutation, always removing the chosen element of ιN before making the next selection.

For example with $N=4$, $I=10$ the 3-factorial representation of I is 1 2 0, and so we start with

 1 2 3 4 along with 1 2 0 .

Now use 1 from 1 2 0 to select 2 as the first element of the permutation leaving

 1 3 4 along with 2 0 .

Now use 2 from 2 0 to select 4 as the second element of the permutation leaving

  1 3 along with 0 .

9. Combinatorics

Use 0 to select 1 as the third element leaving 3 as the final element to give the required permutation as 2 4 1 3.

If R is a factorial number and L is a vector to be permuted which may be any vector (character or numeric) of length at least one greater than ρR, this algorithm may be expressed as

$$LEXPERM: \; L[T],((L[T\leftarrow1+1\uparrow R]\neq L)/L)$$
$$LEXPERM \; 1\downarrow R \; : \; 0=\rho R \; : \; L$$

Examples :

```
      1 2 3 4 LEXPERM 1 2 0
2 4 1 3
      'ABCD' LEXPERM 1 2 0
BDAC
```

This may now be combined with FACTINT to give a program which returns the Rth permutation of the vector ιL.

$$PERM \; : \; (\iota L)LEXPERM \; (L-1)FACTINT \; R-1$$

so that we have for example

```
      4 PERM 21
4 2 1 3
```

A complete list of the first R permutations of order L is given by

$$PERMS \; : \; (L \; PERMS \; R-1),[1]L \; PERM \; R \; :$$
$$R=0 \; : \; (0,L)\rho 0$$

Example :

          3 PERMS 6
    1 2 3
    1 3 2
    2 1 3
    2 3 1
    3 1 2
    3 2 1

A permutation R can be converted into a binary form by

          BINPERM : (ιρR)∘.=R

Examples:

          3 PERM 4
    2 3 1
          BINPERM 3 PERM 4
    0 0 1
    1 0 0
    0 1 0

i.e. the 1's in the <u>columns</u> are located in rows 2,3 and 1 respectively.

9.2. Derangements

A derangement is a permutation in which none of the items are in their natural place. Derangements of order L can therefore be selected from permutations by modifying **PERMS** to:

          DERANGE : (L DERANGE R-1),[1]
                    ((~∨/T=ιL),L)ρT←L PERM R :
                                    R=0 : (0,L)ρ0

9. Combinatorics

The complete list of derangements of order 3, for example, is

```
      3 DERANGE 6
2 3 1
3 1 2
```

The number of derangements of a given order is

$$NUMDER : (!R) \times -/ \div !0, \iota R$$

I. R people each pick up a hat from a cloakroom in which the lights have gone out. How many people have a wrong hat? Simulate this situation with

$$HATS : +/(\iota R) \neq R?R$$

Repeat this experiment L times and find the total number of instances of a wrong hat:

$$RPTHATS:(HATS\ R)+(L-1)RPTHATS\ R: L=0: 0$$

You should observe that the average number of people with a wrong hat tends to R-1, e.g.

```
      (100 RPTHATS 6)÷100
5.1
```

Now investigate the proportion of experiments in which there is the hats are totally mismatched by changing $(HATS\ R)$ to $(R=HATS\ R)$ in $RPTHATS$:

```
      (100 RPTHATS 6)÷100
0.34
```

This result should tend to $1/e = 0.3679$ as L and R increase.

9.3. Combinations

To obtain all combinations of L integers out of R, we first revisit the function SOMESET in Chapter 3 which returns a matrix whose columns represent combinations in the form of binary integers.

```
      2 SOMESET 4
0 0 0 1 1 1
0 1 1 0 0 1
1 0 1 0 1 0
1 1 0 1 0 0
```

The first column represents the combination 3 4 since the 1's are in rows 3 and 4, and similarly for the other columns.

The function BTONUM carries out this transformation recursively column by column:

$$BTONUM: \ (R[;1]/\iota 1\uparrow\rho R),BTONUM \ 0 \ 1\downarrow R \ :$$
$$1=^{-}1\uparrow\rho R \ :(,R)\neq(\rho R)\rho\iota\rho,R$$

Example :

```
      BTONUM 2 SOMESET 4
3 2 2 1 1 1
4 4 3 4 3 2
```

A function to enumerate as rows all the combinations of L objects out of R is thus

$$COMBS \ : \ \lozenge BTONUM \ L \ SOMESET \ R$$

9. Combinatorics   141

Example :
          2 *COMBS* 4
3 4
2 4
2 3
1 4
1 3
1 2

9.4. Selections

An algorithm for all selections of L integers out of R, with repetitions demonstrates a further application of the encode function. Note that the number of such selections is R^L. The selections themselves are the columns of the result.

$$SELECT : 1+(L\rho R)\top {}^{-}1+\iota R*L$$

```
         2 SELECT 3                    3 SELECT 2
1 1 1 2 2 2 3 3 3            1 1 1 1 2 2 2 2
1 2 3 1 2 3 1 2 3            1 1 2 2 1 1 2 2
                             1 2 1 2 1 2 1 2
```

A list of all 3-letter words which can be made from the alphabet 'AB' is

         '*AB*'[3 *SELECT* 2]
AAAABBBB
AABBAABB
ABABABAB

If the selections are regarded as internally unordered use

         ⍒*REMDUPM RSORTU* ⍒2 *SELECT* 3
1 1 1 2 2 3
1 2 3 2 3 3
(see Appendix 2 for REMDUPM and RSORTU)

The number of such selections is $L!L+R-1$. The fact that this is a number of combinations means that the above list of numbers could also have been derived from 2 SOMESET 4 (see previous page). Within each column interpret the 0's as "dividers" betweens 1's, 2's, 3's,. Thus column 1 with 2 dividers at the start represents no 1's, no 2's, and two 3's. Column 2 is no 1's, one 2, one 3, and so on. The function BTONUM can be amended to implement this:

```
BTONUM1:(1+R[;1]/+\~R[;1]),
       BTONUM1 0 1↓R : 1=¯1↑ρR :
                   1+(,R)⊤+\~R
```

giving

```
        BTONUM1 2 SOMESET 4
3 2 2 1 1 1
3 3 2 3 2 1
```

9.5. Compositions and Partitions

A composition of an integer L is an allocation of L indistinguishable items to an unspecified number of non-empty ordered cells. The compositions of 4 are thus

3 1
2 2
2 1 1
1 3
1 2 1
1 1 2
1 1 1 1
4

A single composition of L is given by

```
     COMPOSN: 1 DIF 0,((((L-1)ρ2)⊤R),1)/ιL
```

9. Combinatorics 143

where R is an integer in the range 1 to 2^{L-1}, and DIF is as given in Chapter 4. A matrix of the first R compositions of L is given by

```
COMPS : (L COMPS R-1),[1]L↑L COMPOSN R :
                   R=0 : (0,L)ρ0
```

To display the compositions one by one as shown above, insert ⎕← before L COMPOSN R.

If compositions such as 2 1 1 and 1 2 1 which differ only in internal ordering are judged to be equivalent, we talk of the <u>partitions</u> of R. Each partition corresponds to a way in which R can be expressed as the sum of integers less than or equal to R. Partitions can be obtained from compositions by

```
    REMDUPM RSORTD COMPS 4
3 1 0 0
2 2 0 0
2 1 1 0
1 1 1 1
4 0 0 0
```

9.6. Latin Squares

Latin squares are combinatorial patterns of letters or numbers which are much used in the Design of Experiments. In the square of order R each row and column contains exactly one occurrence of the integers 0,1,...(R-1). The number of distinct Latin squares of order R grows rapidly as R increases — the program below gives the simplest pattern which can be obtained for all integers greater than 1.

```
LATIN : R|(⌽T)∘.-T←ιR
```

```
      LATIN 4
3 2 1 0
2 1 0 3
1 0 3 2
0 3 2 1
```

9.7. Magic Squares of Odd Order

Magic squares are patterns of numbers which belong mainly to the world of recreational mathematics. There are many variations on the basic theme which is that all rows, columns and two major diagonals add up to the same total. As with Latin squares, we give a simple algorithm which produces a magic square of any odd order 2R + 1.

$$MAGIC\ :\ T \ominus (-\phi T \leftarrow \iota R) \phi (\lfloor .5 \times R) \ominus (2 \rho R) \rho \iota R \star 2$$

```
        MAGIC 5
17 24  1  8 15
23  5  7 14 16
 4  6 13 20 22
10 12 19 21  3
11 18 25  2  9
```

Chapter 10

Games and Miscellaneous

10.1. Deal a Hand at Whist

The following program to deal whist/bridge hands exhibits yet another use of encode, this time to assign a random drawing from the 52 cards of a pack to a suit/value pair. The assignment is made numerically in line 1, and card values are assigned by indexing in line 6 which also sorts each hand in suit order.

```
     ∇ HAND;T;I;J;U
[1]    T←1+0 13⊤4 13ρ⁻1+52?52
[2]    I←0
[3]  L1:→0 IF 4<I←I+1
[4]    J←0 & ''
[5]  L2:→L1 IF 4<J←J+1
[6]    'SCHD'[J],'-','23456789TJQKA'
            [U[⍋U←(J=T[1;I;])/T[2;I;]]]
[7]    →L2
     ∇
```

Example :

   HAND

S–QT3
C–T3
H–AJ872
D–K64

S–J7
C–A9842
H–T5
D–Q872

S–A9862
C–K6
H–K864
D–A9

S–K54
C–QJ75
H–93
D–JT53

10.2. Chessboard

This program returns a chessboard pattern made up of the characters '*' and 'blank'.

$$CHBOARD: \ ' * \ '[1+(CHVECT \ R[1]) \circ . = CHVECT \ R[2]]$$

R : 2-item vector giving size in characters of one of the 64 squares of the board

10. Games and Miscellaneous

The auxiliary function CHVECT returns alternate sequences of 1's and 0's, each of length R.

$$CHVECT : 2|\lfloor(^{-}1+\iota 8\times R)\div R$$

10.3. Mastermind

The six colours of the standard Mastermind game are represented as the first six positive integers. The computer sets a pattern (line 1), and alternately prompts and scores the user until a correct pattern match is achieved. A "black" is a perfect match, colour and position, a "white" is a match for colour only.

```
        ∇ Z←MM;T;U
[1]     Z←4?6
[2]     L1:T←ASK'YOUR GUESS?'
[3]     →L2 IF ∧/T=Z
[4]     'WHITES = ',(⍕0⌈((+/T∊Z)⌊+/Z∊T)-U),
                      ' BLACKS = ',⍕U←+/T=Z
[5]     →L1
[6]     L2:'WELL DONE - CORRECT ANS. IS :-'
        ∇
```

Example :

```
        MM
YOUR GUESS?
2 3 4 5
WHITES = 2, BLACKS = 1     etc.
```

10.4. Life

The game of Life is played on a rectangular board whose initial dimensions are provided by the player in line 2. A set of initial occupied cells is determined by the player giving a series of row and column co-ordinates terminating with a null line (lines 3 and 4). Then the number of generations required must be given (line 6).

The rules of Life are embodied in line 10, namely that at each new generation, each cell, whether occupied or not makes a count of its neighbours. A "neighbour" is an adjoining cell – horizontally, vertically or diagonally – so that each cell on the board has 8 neighbours – this is reflected in the 8 subexpressions in lines 8 and 9. Any cell with 3 neighbours becomes or remains occupied, while an <u>occupied</u> cell with 2 neighbours also remains occupied. (Occupied cells with more than 3 neighbours are assumed to die from overcrowding, those with less than 2 to die from boredom!).

```
         ∇ LIFE;T;T1;U
[1]      T←BORDER(T←ASK'SIZE OF BOARD?')ρ0
[2]      'STARTING FIGURE '
[3]      L1:→L2 IF 0=ρU←ASK'NEXT CELL?'
[4]      →L1,T[U[1];U[2]]←1
[5]      L2:' *'[1+T]
[6]      →0 IF 0=ρU←ASK'NO OF GENERATIONS?'
[7]      L3:→L2 IF 0>U←U-1
[8]      T1←(1⊖T)+(1⌽¯1⊖T)+(1⌽T)+1⌽1⊖T
[9]      T1←T1+(¯1⌽¯1⊖T)+(¯1⌽T)+(¯1⌽1⊖T)+¯1⊖T
[10]     T←(T1=3)∨(T1=2)∧T=1
[11]     →L3
         ∇
```

10. Games and Miscellaneous

For BORDER see Appendix 2. The current state of the board is displayed as a character matrix in line 5. This game became the subject of an extensive literature a few years ago, see e.g. Martin Gardner's articles in the Scientific American Vol 223 no. 10 (Oct. 1970) and Vol 224 No. 2 (Feb 1971).

10.5. Recursive Algorithms

10.5.1. Tower of Hanoi

At the start L rings, all of different sizes, are piled on the leftmost of 3 pegs (peg 1) with the rings in increasing order of size going down the peg. The object is to repeat this configuration on the rightmost peg, moving only one ring at a time, and without ever allowing a ring to rest on a ring smaller than itself. The following program achieves this with the minimum number of moves.

```
     ∇ L HANOI R
[1]    →0 IF L=0
[2]    (L-1)HANOI R[1 3 2]
[3]    (⍕L),':',(⍕R[1]),'→',⍕R[2]
[4]    (L-1)HANOI R[3 2 1]
     ∇
```

    L : number of pegs to be moved
    R : vector of numbers representing source,
      target and spare pegs respectively
      (say 1 2 3)

The principle is that once the first N-1 rings have been successfully transferred to the spare peg (line 2), the Nth. ring is then transferred to the target peg (line 3), and the first N-1 rings are then moved on top of it (line 4).

10.5.2. Ackerman's Function

If you have not met it before, this innocent-looking function will amaze you by its capacity to gobble up computer time in evaluating the function for some small values of its (positive integer or zero) arguments. The function is defined first in algebra, and then in APL :

$$f(0,R) = R+1$$
$$f(L,0) = f((L-1),1)$$
$$f(L,R) = f((L-1),f(L,R-1))$$

```
       ∇ Z←L ACKER R
[1]    →L2 IF ~0∈L,R
[2]    →L1 IF L≠0
[3]    →0,Z←R+1
[4]    L1:→0,Z←(L-1)ACKER 1
[5]    L2:Z←(L-1)ACKER L ACKER R-1
       ∇
```

Example :

```
       3 ACKER 1
13
```

10.6. Optical Illusions

Some well known optical illusions are easy to effect on the screen of a small micro. Here are three of the best known simple illusions - realising them on a computer gives the opportunity to experiment with variations on a basic theme. The defined functions used are described in Appendix 1.

10. Games and Miscellaneous   151

Which small square has the larger area?

0 0 511 255 *FILLBOX* 3 5 ρ100 100 200 200 1
                             125 125 175 175 0
                             325 125 375 175 1

0 0 511 255 *BOXES* 1 4ρ 300 100 400 200

Which horizontal line is the longer?

        A←2 *REFLECT* 1 *REFLECT* 3 3ρ0 0 0
                                     1 2 0
                                     1 1 1
        B←2 *REFLECT* 1 *REFLECT* 3 3ρ0 0 0
                                     1 2 0
                                     1 3 1
        A[;2]←A[;2]−4
        B[;2]←B[;2]+4
        ¯10 ¯5 10 5 *SKETCH* A,[1]B

Which circle is the largest?

        T←3 3ρ 0 55 90
               1 10 50
               1 55 10
        Y←3 3ρ 30 50 10
               50 50 10
               70 50 10
        W←0 0 120 120

Use *W SKETCH T* & *W CIRCLES Y* then clear the screen and try *W CIRCLES Y* alone.

Appendix 1

Graphics

There are many points in the preceding chapters, particularly the one in geometry, where the exposition of mathematics is greatly enhanced by a <u>display</u> of the results of running a program.

APL itself contains no graphics primitives and so the production of good pictures depends on the availability of some additional resource − normally a separate graphics package − whereby arrays can be rendered into pictures.

It is impossible to anticipate the range of graphics packages which might be used in conjunction with APL in the classroom, so it has been assumed that a basic package is available containing a number of primitive graphic functions which are described below.

A common underlying concept is that of a "window." Conceptually there are two sets of co-ordinates which the user must keep in view, graphics co-ordinates and problem-space co-ordinates. The former are determined by the physical nature of the equipment and are intimated in the appropriate manual. To be specific, suppose that the co-ordinates of a screen boundary are

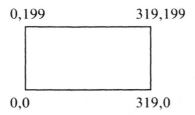

The problem-space co-ordinates on the other hand are determined entirely by the user, who perceives the screen boundary as determined by the extreme values of x and y which he/she wishes to be represented on the screen.

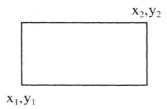

If the user co-ordinates of the bottom left and top right corners of the screen are (x_1,y_1) and (x_2,y_2) then the vector

$$x_1,y_1,x_2,y_2$$

is called the "window."

A fundamental auxiliary function is XFM which transforms a matrix of values in problem-space co-ordinates into the corresponding values in graphics co-ordinates. The left argument of XFM is the window, the right argument consists of a single column of x-co-ordinates, followed by one or more columns which are interpreted as y-values. Given the screen dimensions above, XFM is

Appendix 1. Graphics

```
      ∇ Z←L XFM R;I
[1]   I←1
[2]   Z←((1↑ρR),1)ρINT 319×(R[;1]-L[1])÷
                                     -/L[3 1]
[3]   L1:→0 IF (¯1↑ρR)<I←I+1
[4]   Z←Z,INT 199×(R[;I]-L[2])÷-/L[4 2]
[5]   →L1
      ∇
```

The function INT rounds a number to the nearest integer (see Appendix 2). The values 319 and 199 are screen boundary co-ordinates (see previous page) and should be replaced by appropriate values for the user's own system. The graphics functions themselves are

    L DRAW R

        L : window
        R : (NxP) data matrix, col 1 = x-values
        other cols. = sets of y-values

Draws (P-1) sets of (N-1) connected line segments - the first given by $R[;1\ 2]$, the next by $R[;1\ 3]$ and so on.

    L PLOT R

As for **DRAW** except that the points are plotted but not connected.

    L PIN R

        L : window
        R : 2 element vector (x,y)

Draws the single point R as a pin-point or pixel.

L JOIN R

        L : window
        R : 2 element vector (x,y)

Plots a point and draws a line to it from the previously plotted position.

L SKETCH R

        L : window
        R : 3 column matrix.
            cols 2,3 are sets of (x,y) co-ordinates respectively;
            elements of col 1 are 0 = move, 1 = draw.

Similar to **DRAW** for a single graph but adding a "penup/pendown" facility.

L LINES R

        L : window
        R : (Nx4) matrix. Each row is a set of 4 co-ordinates in the form (x_1, y_1, x_2, y_2) representing the start and end points of a line.

Draws as many lines as there are rows of R.

L BOXES R

        L : window
        R : (Nx4) matrix. Each row is a set of 4 co-ordinates in the form (x_1, y_1, x_2, y_2) representing the lower left and upper right corners of a box.

Outlines as many rectangles as there are rows of R.

Appendix 1. Graphics

L FILLBOX R

> L : window
> R : (Nx5) matrix. First 4 columns as for BOXES. 5th. column is an integer representing the colour with which the box is to be filled. 0 = no fill.

Fills as many rectangles as there are rows of R.

L ARCS R

> L : window
> R : (Nx5) matrix. Each row is a set of 4 co-ordinates in the form (x_1, y_1, x_2, y_2) representing the end-points of an arc, followed by the angle in degrees of the arc connecting them.

Draws as many arcs as are defined by the rows of R.

L CIRCLES R

> L : window
> R : (Nx3) matrix. Each row is (x,y,r) where (x,y) are the co-ordinates of the centre and r is the radius.

Draws as many circles as are defined by the rows of R.

HIST R

> R : vector

Draws a histogram the heights of whose boxes are the elements of R. The histogram is automatically scaled so that the picture fills the screen.

In addition it is useful to have a function REFLECT which for a matrix R, negates the values in the column(s) L, and joins the result to the original matrix

    REFLECT: $Z \leftarrow R$ &
        $Z[;L] \leftarrow -Z[;L]$ & $Z \leftarrow R,[1]Z$

Example :

```
      R←2 2ρ3 4 5 6
      1 REFLECT R              2 REFLECT R
3 4                         3  4
5 6                         5  6
¯3 4                        3 ¯4
¯5 6                        5 ¯6
```

This set of functions covers all the graphics needs of the algorithms described in this book. The user must code them using the system functions available in the APL/graphics system to hand, perhaps adapting them to a set of functions which fit more naturally within that system.

It should be added that if no graphics system is available, simple graphics may nevertheless be produced on a character-box basis as in the function LIFE (line 8) in Chapter 10. For example, the sequence

Appendix 1. Graphics

```
      X←ι6
    ' *'[1+(⌽¯2+ι10)∘.=⌊'(X-2)×X-4']
*



*    *


  *  *
    *
```

draws a graph of the parabola y = (x-2)(x-4) from x = 1 to 6.

Appendix 2

Idioms and Utilities

This section contains a collection of general purpose programs relevant to mathematics. Some are useful in their "stand-alone" role - others occur as auxiliary functions in programs throughout this book. Many of the phrases given here have become part of the folk-lore of APL, and are recognised almost like primitives by habitual users.

A2.1. Rounding, Averaging, and Removing Duplicates

Round number(s) R to L decimal places

$ROUND$: $(10*-L)\times\lfloor.5+R\times10*L$

L : integer (possibly 0 or negative - if negative
R is rounded to nearest $10*-L$
R : numeric scalar, vector or array

Special case of the above - round R to nearest integer

$INT\ :\ \lfloor.5+R$

Average of a vector R

$AVERAGE\ :\ (+/R)\div\rho R$

Remove duplicate elements from a vector

$REMDUP\ :\ ((R\iota R)=\iota\rho R)/R$

alternatively

$REMDUP\ :\ (\vee/<\backslash R\circ.=R)/R$

Remove duplicate rows from a matrix

$REMDUPM:\ (1\ 1\phi<\backslash R\wedge.=\transpose R)\neq R$

A2.2. Sorting and Ranking

Sort a vector R in upward/downward sequence

$SORTUP\ :\ R[\blacktriangle R]\qquad SORTDWN\ :\ R[\blacktriangledown R]$

Obtain upward/downward ranks for vector R

$RANKUP\ :\ \blacktriangle\blacktriangle R\qquad RANKDWN\ :\ \blacktriangle\blacktriangledown R$

Appendix 2. Idioms and Utilities

Obtain upward tied ranks for vector R

$TRANKU$: $.5\times(\blacktriangle\blacktriangle R)+\blacktriangledown\blacktriangle\phi R$

Obtain downward tied ranks for vector R

$TRANKD$: $.5\times(\blacktriangle\blacktriangledown R)+\blacktriangledown\blacktriangledown\phi R$

Obtain downward rank for vector R with equal members being given lowest possible rank (schoolmaster's rank) :

$SRANK$: $(\blacktriangle\blacktriangledown R)[R\iota R]$

Change $\blacktriangledown$ to $\blacktriangle$ for upward schoolmaster's rank.

Examples :

```
      MARKS
37 85 22 37 37 22
      TRANKU MARKS
4 6 1.5 4 4 1.5
      TRANKD MARKS
3 1 5.5 3 3 5.5
      SRANK MARKS
2 1 5 2 2 5
```

Merge vector R into vector L with items in correct order

$MERGE$: $(L,R)[\blacktriangle L,R]$

Example :

```
      0 10 20 30 MERGE 27 13 32
0 10 13 20 27 30 32
```

Sort each row of a matrix in ascending order

$$RSORTU \ : \ T\rho(R[\triangle R])[\triangle\lceil(\triangle R\leftarrow,R)\div{}^{-}1\uparrow T\leftarrow\rho R]$$

Sort each row of a matrix in descending order

$$RSORTD \ : \ T\rho(R[\triangledown R])[\triangle\lceil(\triangledown R\leftarrow,R)\div{}^{-}1\uparrow T\leftarrow\rho R]$$

Example :

```
      M
7  7  10
2  8  5
6  3  1
     RSORTU M
7  7  10
2  5  8
1  3  6
```

Sort (a) rows (b) columns of a matrix in ascending order

$$UPMATR \ : \ R[\triangle\lceil(\lceil/,R)\bot\lozenge R;]$$

$$UPMATC \ : \ R[;\triangle\lceil(\lceil/,R)\bot R]$$

n.b. leftmost column/topmost row is most significant within row/column

Example :

```
     UPMATR M
2  8  5
6  3  1
7  7  10
```

This technique can be extended to sort the rows of a character matrix in alphabetic order :

Appendix 2. Idioms and Utilities   165

         UPSORT : ⎕AV[UPMATR ⎕AV⍳R]

Example :

         MC
SANTA
CLAUS
ESQ
         UPSORT MC
CLAUS
ESQ
SANTA

A2.3. Statement Joining

"Glue"

In the introduction the use of the '&' symbol was described as a means of keeping two related statements together on the same line. In some implementations this is acceptable as a correct APL construct; in others one must either write two separate statement lines, or else use a device such as the following to "glue" two statements together

         ...,0⍴...

This is a technique which is sometimes frowned on by APL purists, because of a trap which can arise from e.g.

         A←2,0⍴B←2

Although apparently identical, B is a scalar whereas A is a 1-element vector. Provided that this possibility is appreciated, the device can be used to compress two simple statements onto one line, and can be useful when the rightmost statement is an output message, e.g.

$A\leftarrow 2,0\rho\square\leftarrow\text{'}PROCESS\ TERMINATES\text{'}$

Merging Branch Statements

Another statement joining technique uses the fact that a branch to a non-null vector means branch to its first item, ignoring all other items. Thus

$\rightarrow L1,Z\leftarrow\ ...$

joins a branching statement to another statement on its right. The result of the other statement must be a <u>numeric</u> scalar or vector - sometimes it may require an extra ravel to achieve this, i.e.

$\rightarrow L1,,Z\leftarrow\ ...$

Read this as "branch to L1, having assigned Z to ...." See SER1 in Chapter 4 for an example of this technique.

Mid-line Assignment

Yet another device which is frequently used to merge two or more statements on one line is mid-line assignment, of which the technique described above is a special case. It is frequently used to increment loop-counting variables (see line 2 of AMICNOS), and another general example from Chapter 2 is

$PRIMES\ :\ (2=+/0=T\circ .|T)/T\leftarrow\iota R$

It is a good idea to read this as

"(...) reduce ιR which is incidentally assigned to T."

Some APL users discourage mid-line assignment on the grounds that in reading the line above from left to right, you cannot "understand" T in the parentheses until you get to a later stage in the APL sentence. While it is true that overuse can lead to

Appendix 2. Idioms and Utilities 167

incomprehensible program lines, its use in moderation can often be justified.

A2.4. Branching and Prompting

Two other useful functions are IF and ASK. The function IF is used repeatedly in this text. ASK is used occasionally to show its possibilities; in classroom practice it can provide user-friendly interaction with some of the mathematical functions. In order to keep the <u>mathematical</u> aspects as clear as possible, such "cosmetics" have been trimmed in the programs given. This is not however to underestimate their importance in the overall presentation of mathematical software.

$$IF\ :\ R/L$$

   L : label
   R : condition

This construction is widely used in this text to clarify conditional branches by writing e.g.

$\rightarrow L1\ IF\ R<I\leftarrow I+1$ rather than $\rightarrow(R<I\leftarrow I+1)/L1$.

Prompting

```
     ∇  Z←ASK R
[1]  R
[2]  →L1 IF 0≠ρZ←⎕
[3]  →0,Z←ι0
[4]  L1:Z←⍎Z
     ∇
```

Example :

   $A\leftarrow,ASK\ 'GIVE\ 2\ NUMBERS:'$

following which A will have the value of whatever APL expression (which includes of course a constant scalar or vector) is given by the user in response to the prompt. See CAST9 in Chapter 2 for a further example.

A2.5. Matrix Manipulation

Column and row addition of vector L to matrix R

       $COLADD$: $R+(\rho R)\rho L$

       $ROWADD$: $\otimes((\phi\rho R)\rho L)+\otimes R$

           L : vector with same number of items as R has cols/rows
           R : numeric matrix

Example :

```
            0 1 2 ROWADD 3 3ριθ
    1  2  3
    5  6  7
    9 10 11
```

Upper and lower triangular RxR matrices

These have 1's in upper/lower triangle (excluding the leading diagonal), 0's elsewhere.

       UPT : $(\iota R)\circ.<\iota R$ and LOT : $(\iota R)\circ.>\iota R$

Ditto including the leading diagonal

       $UPTD$: $(\iota R)\circ.\leq\iota R$ and $LOTD$: $(\iota R)\circ.\geq\iota R$

Appendix 2. Idioms and Utilities

```
1 1 1
0 1 1
0 0 1
```

Unit matrix of order R

$$UNIT \ : (\iota R)\circ .=\iota R \quad \text{or} \quad UNIT \ : (R,R)\rho(R+1)\uparrow 1$$

Border a numeric matrix with zeros

$$BORDER \ : \ (-2+\rho R)\uparrow(1+\rho R)\uparrow R$$

A2.6. Replication

The following function implements the function **REPLICATE** which is to be found in some APL's as an extension of COMPRESS. R is a vector (character or numeric), and L is a numeric vector of the same length indicating for each item of R how many times (possibly zero) it is to be replicated. COMPRESS is then a special case in which only 0's and 1's are allowed in L.

```
       ∇ Z←L REPLIC R
[1]    Z←ι0
[2]    L1:→0 IF 0=ρL
[3]    Z←Z,(1↑L)ρ1↑R
[4]    L←1↓L  &  R←1↓R
[5]    →L1
       ∇
```

Example :

```
      2 3 0 1 REPLIC 'ABCD'
AABBBD
```

A2.7. Without

Another function which is to be found in some APLs as dyadic "not" is WITHOUT, which is approximately modelled by

$$WITHOUT \; : \; (\sim L \epsilon R)/L$$

Example :

```
      (ι10)WITHOUT 4 5 6
1 2 3 7 8 9 10
```

A2.8. Bit Manipulation

In the following set of idioms involving the scan operator, a bit is considered to be switched "on" if it has the value 1, "off" if it has the value 0.

Switch on all bits from and after the first 1   : $\vee \backslash R$

Switch on the first 1, off all bits thereafter   : $< \backslash R$

Switch on all bits after the first 0   : $\leq \backslash R$

Switch off all bits after the first 0   : $\wedge \backslash R$

Count the number of 1's before the first 0   : $+/\wedge \backslash R$

Examples :
```
      M1
0 0 0 0 1
0 1 1 1 1
0 1 0 1 0
1 0 1 0 1
1 0 0 0 0
1 1 1 1 0
```

```
        ∨\M1                    ∧\M1
    0 0 0 0 1              0 0 0 0 0
    0 1 1 1 1              0 0 0 0 0
    0 1 1 1 1              0 0 0 0 0
    1 1 1 1 1              1 0 0 0 0
    1 1 1 1 1              1 0 0 0 0
    1 1 1 1 1              1 1 1 1 0

        <\M1                    ≤\M1
    0 0 0 0 1              0 1 1 1 1
    0 1 0 0 0              0 1 1 1 1
    0 1 0 0 0              0 1 1 1 1
    1 0 0 0 0              1 0 1 1 1
    1 0 0 0 0              1 0 1 1 1
    1 0 0 0 0              1 1 1 1 0
```

A2.9. Some String Handling Functions

Although mathematics makes less demands than most other application areas on the string handling capabilities of APL, these are occasionally useful, and so three general utilities are given here.

Blank out all occurrences of a specified set of characters

$$BLANK \; : \; T\backslash(T \leftarrow \sim L \in R)/L$$

        L : string in which replacement is to take place
        R : string of characters to be replaced by blanks

Example :

```
     '(2X-3Y).(4X+5Y)' BLANK 'XY().'
  2 -3    4 +5
```

Find all occurrences of one string in another

$$FINDALL: \; R\wedge.=(^{-}1+\iota\rho R)\phi((\rho R),\rho L)\rho L$$

L : string to search in
R : string to search for

The result of FINDALL is a bit vector of the same length as L, with 1's marking the start position of matches of R within L.

Example :

`'(2X-3Y).(4X+5Y).(X-7Y)' FINDALL ').('`
`0 0 0 0 0 0 1 0 0 0 0 0 0 0 1 0 0 0 0 0 0 0`

A more general utility which covers many of the circumstances where string handling is useful in a mathematical context is REPLACE which replaces all occurrences of one string in L with another.

```
      ∇ Z←L REPLACE R;T;T1;T2;U
[1]   T1←(T←R⍳';')↓R & T2←(T-1)↑R
[2]   Z←'' & U←L FINDALL T2
[3]   L1:→L2 IF 0=∨/U
[4]   Z←Z,((T←⁻1+U⍳1)↑L),T1
[5]   L←(T←T+ρT2)↓L & U←T↓U
[6]   →L1
[7]   L2:Z←Z,L
      ∇
```

L : string in which replacements are to be made
R : string to be replaced,
    followed by semi-colon,
    followed by replacing string

Appendix 2. Idioms and Utilities   173

Line 1 separates out the replaced and replacing strings; line 2 establishes the points where replacement is to take place, and the rest of the function is a loop to do the replacements one by one. If ';' is one of the characters which has to be replaced, a different separator character must be used, and an appropriate adjustment made to line 1.

Examples :

```
      '2 -3  4 +5'  REPLACE  '-;⁻'
2 ⁻3  4 +5
      '2,399,999'  REPLACE  '99;00'
2,300,009
      '2,399,999'  REPLACE  ',;'
2399999
```

A2.10. Testing for Numeric/Character

The function TESTNUM returns 1 if R is numeric and 0 if R is character

$$TESTNUM \; : \; 0=1\uparrow 0\rho R$$

As noted earlier, the programs in this collection have been stripped of data-checks in order to help clarify essential program structure. For versions intended for routine use, and which might be passed on to other users, it is wise to commence each program with a prologue which does routine checking. For example, if both arguments of a function must be numeric, L must be a scalar and R a matrix, begin the function

[1]   →L0 IF (TESTNUM L)∧(TESTNUM R)∧
                            (0=ρρL)∧2=ρρR
[2]   →0,0ρ☐←'ERROR'
[3]   L0: ... function proper

A2.11. Timing Function Execution

One utility function which most users eventually want is a TIMER function. The exact contents depend on how the user's own system reports CPU time used. Here is a TIMER which assumes that current CPU time is recorded in $\Box AI[2]$, and that R is the character string form of a phrase which it is required to time for L executions

```
       ∇  Z←L TIMER R;I
[1]    I←0 & Z←⎕AI[2]
[2]    L1:→L2 IF L<I←I+1
[3]    ⍎R
[6]    →L1
[7]    L2:Z←⎕AI[2]-Z
       ∇
```

Example (on one specific computer) :

```
       100 TIMER 'X←?100⍴100'
86
```

Appendix 3

Graphics Functions in I-APL

Using I-APL and the supplied workspace PGRAPH, the following functions implement the graphics functions of Appendix 1 on IBM PC's and compatible machines. Direct definitions should be entered as given, i.e. α and ω correspond to L and R elsewhere in the book.

$DRAW:(\alpha DRAW\ 0\ ^{-}1\downarrow\omega)$,
　　$(32767\ 32767\ 0\ ^{-}1\ 0\ 0\ 0\ 0,^{-}1+1\uparrow\rho\omega)$
　　　$PGPLOT\lfloor\alpha XFM\omega[;1,T]\ :\ 1=T\leftarrow^{-}1\uparrow\rho\omega\ :\iota 0$

$PLOT:(\alpha PLOT\ 0\ ^{-}1\downarrow\omega)$,
　　　$PGPLOT\lfloor\alpha XFM\omega[;1,T]\ :\ 1=T\leftarrow^{-}1\uparrow\rho\omega\ :\iota 0$

$PIN:PGPLOT\alpha XFM1\ 2\rho\omega$

　　　$\nabla\ L\ SKETCH\ R;T$
[1]　　$L1:\rightarrow 0\ IF\ 0\geq 1\uparrow\rho R$
[2]　　$R[1;1]\leftarrow 1$
[3]　　$T\leftarrow^{-}1+R[;1]\iota 0$
[4]　　$L\ DRAW\ R[\iota T;2\ 3]$

[5]   R←(T,0)↓R
[6]   →L1
        ∇

The **PGRAPH** functions have no concept of "current graphics pointer," and so the function **JOIN** must be programmed separately for each application, with a variable recording the coordinates of the last plotted point. The functions for **LINES**, **BOXES**, and **FILLBOX** are

$$LINES: 32767\ 32767\ 0\ ^-1\ 0\ 0\ ^-1\ ^-1\ 1\rho\omega$$
$$PGPLOT\lfloor\alpha XFM(2\ .5\times\rho\omega)\rho\omega$$

        ∇ L BOXES R
[1]   L1:→0 IF 0≥1↑ρR
[2]   L DRAW R[1;3 3 1 1 3],[1.5]
                              R[1;2 4 4 2 2]
[3]   R←1 0↓R
[4]   →L1
        ∇

        ∇ L FILLBOX R;T;U
[1]   L1:→0 IF 0≥1↑ρR
[2]   T←⌊L XFM 2 2ρR[1;]
[3]   U←|-/T[;2]
[4]   T[2;2]←T[1;2]
[5]   (32767 32767 0 ¯1 0,U,0 0 1)PGPLOT T
[6]   →L1
        ∇

The function to implement circles assumes that each circle is drawn as an 18-sided polygon. If higher resolution is required, change the '19's' in line 2 to one more than the number of sides in the approximating polygon.

Appendix 3. Graphics Functions in I-APL

```
     ∇ L CIRCLES R
[1]  L1:→0 IF 0≥1↑ρR
[2]  L DRAW (19 2ρR[1;⍳2])+
              R[1;3] CIRC 20×⍳19
[3]  R←1 0↓R
[4]  →L1
     ∇
```

For CIRC see Section 6.1.

ARCS defines arcs as a series of connected line segments. The function ARC embodies the necessary co-ordinate geometry to give the anti-clockwise arc of angle $R[5]$ degrees connecting the points $R[1\ 2]$ and $R[3\ 4]$. To obtain the clockwise arc reverse the pair of co-ordinates.

The function ARC assumes that points will be connected at 5 degree intervals. If greater or lesser resolution is required, change the value 5 in line 2 of ARCS to the desired value.

The auxiliary function PTS requires for R an anti-clockwise range of degrees as a 2-item vector, and for L the resolution in degrees as described above.

```
     ∇ L ARCS R
[1]  L1:→0 IF 0≥1↑ρR
[2]  L DRAW 5 ARC R[1;]
[3]  R←1 0↓R
[4]  →L1
     ∇
```

```
        ∇ Z←L ARC R;T;U;V;W;X;Y
[1]     U←.5×(○¯1↑R)÷180
[2]     T←-/Y←2 2ρ4↑R
[3]     W←((+/T*2)*.5)÷2×1○U
[4]     →L1 IF 0≠1↑T
[5]     V←○.5
[6]     →L2,0ρX←-×1↓T
[7]     L1:V←¯30÷/⌽T
[8]     X←-×1↑T
[9]     L2:Z←(AV Y)+(X×W×2○U)×1 ¯1×1 2○V
[10]    Z←((ρT)ρZ)+T←W CIRC L PTS(U-2×¯1↑T),
               U←(180+X×90)++/T←(180÷○1)×V,U
        ∇


        ∇ Z←PTS R
[1]     R←R+0,360×>/R
[2]     Z←L×(⌊R[1]÷L)+ι⌊(-/⌽R)÷L
[3]     Z←360|R[1],Z,R[2]×ιR[2]≠¯1↑Z
        ∇

        AV:(+/ω)÷1↑ρω
```

The following function implements HIST in Appendix 1:

```
HIST:(0 0 0 ¯1,T,32767,(T←⌊600÷1+ρω),
      0 0,ρω)PGPLOT((ρω),1)ρ⌊ω×180÷⌈/ω
```

Index of Topics

Ackerman's function, 150
Adaptive Simpson's rule, 90, 129-130
Addition of polynomials, 32
Amicable numbers, 14
Angle between lines, 82
Area of a triangle, 59
Arabic numbers, 18
Area of circle, 55
Argument of vector, 81
Arithmetic progressions, 33
Astroid, 71
Average, 103,162
Axis of graph, 8
Backward counting, 23
Best fitting poly, 113
Beta function, 126
Binary fractions, 17
Binomial coefficients, 43
Binomial distribution, 99
Birthday problem, 102
Bisection method, 91
Bit manipulation, 170
Bordering matrix, 169

Box-Muller formula, 106
Branching, 167
Buffon's needle, 108
Cardioid, 75,76
Cartesian co-ordinates, 82ff.
Casting out nines, 21
Centigrade degrees, 55
Chessboard, 146
Chi-squared distribution, 129
Chi-squared test, 125
Circle, 67
Coins, tossing, 108
Combinations, 140
Complex numbers, 24
Complex roots of unity, 25
Compositions, 142
Compound interest, 56
Computer arithmetic, 22
Conditional branching, 167
Conic sections, 67,72
Continued fractions, 51
Co-ordinate geometry, 81
Convergence, 42
Correlation, 115ff.

Cos formula, 60
Covariance, 115ff
Cube, 79
De Moivre's theorem, 35
Decoding, 19
Deltoid, 76
Derangements, 138
Descriptive statistics, 103ff
Determinant, 31
Dice, throwing, 108
Difference of squares, 27
Differences of series, 45
Differential equations, 96
Differentiation, 32
Digit sum, 21
Direct definition, 4
Duplicate removal, 162
E, series for, 50
Ellipse, 67,73
Encoding, 19
Envelopes, 71
Epicycloid, 68,75
Euclid's algorithm, 15
Euclidean norm, 62
Euler's series, 47
Euler's method, 96
Even integers, 7
Expansion of brackets, 28
Exponential distribution, 106
F distribution, 130
Factors, 13
Fahrenheit degrees, 55
Fibonacci series, 40, 46
Finding substrings, 172
Formulae, 55ff.
Forward counting, 23
Fractional parts, 19
Frequency distribution, 109
Gamma distribution, 129
Geometric progressions, 33

Glue, 165
Goodness of Fit, 125
Graphics co-ordinates, 153
Gregory's series, 47
Hastings formula, 127
Highest common factor, 15
Histogram, 157
Hyperbola, 67,73
Hypergeometric distribution, 101
Hypocycloid, 68,75
I-APL Graphics Fns, 175
Idioms, 161ff.
Images, 81
Integration, 32,87
Interpolation, 52
Intersection, 34
Inverse of matrix, 30
Isomorphisms, 11
Iteration method, 92
Joining statements, 165
Kendall coefficient, 119
Koehler's formula, 128
L-norm, 63
Lagrange's formula, 52
Latin squares, 143
Least squares fitting, 113
Lexical ordering, 135
Life, game of, 148
Limacon, 76
Logarithm tables, 11
Longest journey, 61
Lower triangular matrix, 168
Lowest common multiple, 16
Magic squares, 144
Mann-Whitney test, 123
Mastermind, 147
Matrix inverse, 30
Matrix manipulation, 168

Index of Topics

Matrix operations, 30
Matrix product, 30
Matrix quotient, 30
Median, 104
Mensuration, 20
Merging vectors, 163
Mid-line assignment, 166
Mid-point method, 97
Mode, 105
Money conversion, 20
Monotonicity, 42
Moran's formula, 127
Mortgage repayment, 58
Multipln of polynomials, 32
Multiplication tables, 10
Negative exponential dist., 106
Nephoid, 75
Neville's algorithm, 53
Newton-Raphson method, 94ff.
Non-parametric tests, 117ff.
Norms, 62ff.
Normal distribution, 126ff.
Number bases, 16
Numerical integration, 87
Odd integers, 7
Optical illusions, 150
Parabola, 68, 74
Parametric equations, 67
Partition values, 104ff.
Partitions, 143
Pascal's triangle, 44
Patterns of points, 84ff.
Pen-up/pen-down, 66
Percentiles, 104
Perfect numbers, 14
Permutations, 135ff.
Perpendicular distance, 81
Perspective drawing, 80
Pi, 47-49
Poisson distribution, 100
Polar co-ordinates, 82ff.
Polynomials, 31, 36
Polynomial interpolation, 53
Positive integers, 7
Precision, 23
Present value, 58
Primes, 12
Probability density, 126ff.
Problem-space co-ordinates, 153
Product moment correlation, 115
Prompting, 167
Pythagoras theorem, 62
Pythagorean triples, 63
Quadratic equations, 29
Random numbers, 106ff.
Range, 105
Rank correlation, 118
Ranking, 162ff.
Recurrence relations, 39ff.
Recurring decimals, 16
Recursion, 5, 46, 149
Regression, 113
Removing duplicates, 162
Replacing strings, 172
Replication, 169
Residuals, 114
Roman numbers, 18
Root finding, 91ff.
Root 2, series for, 50
Roots of polynomials, 36
Roots of unity, 25
Rotation, 77ff.
Rounding, 161
Runs test, 117

Sample sizes, 131
Scaling axes, 8
Scatterplot, 112
Schoolmaster's rank, 163
Selections, 141
Series, 39ff.
Sets, 34
Shortest journey, 61
Sign test, 120
Simpson's rule, 89
Simulation, 108ff.
Simultaneous equations, 30
Sin formula, 60
Sliding ladder curve, 71
Sorting, 162ff.
Spearman's coefficient, 118
Square numbers, 8
Standard deviation, 104
Statistical tables, 126ff.
Stem and Leaf Plot, 110
String handling, 171ff.
Student's t distribution, 128
Subsets, 35
Sum of squares, 28
Supersets, 35
Surface area of cone, 56
Switching bits, 170
T distribution, 128
Table lookup, 10
Taxi-cab metric, 63

Test for numeric, 173
Tied rank, 163
Timing function, 174
Tower of Hanoi, 149
Transformations, 77
Trapezium method, 97
Trapezium rule, 89
Triangle formulae, 59
Triangular matrices, 168
Triangular numbers, 8
Trigonometric functions, 51
Trigonometric tables, 11
Two-way frequency, 111
Uniform distribution, 106
Union, 34
Unit matrix, 169
Upper triangular matrix, 168
U-test, 123
Utilities, 161ff.
Variance, 103
Velocity, 56
Volume of cylinder, 56
Volume of sphere, 55
Whist, 145
Wilcoxon Test, 122
Wilson-Hilferty approximation, 129
Window, 153
W-test, 122
Yates' correction, 125

Index of Topics

Index of Programs and Variables

ACKER 150	BINPERM 138	COMPOSN 143
ADSIM 91	BINPROB 99	COMPS 143
ALLS 35,36	BISECT 92	CONFRAC 51
AMICNOS 15	BLANK 172	CONIC 72
ANG 82	BORDER 169	CORM 115
ANGLEA 59	BOXES 150,156,176	COS 51
ANTILOG 11	BTONUM 140	COVM 115
AP 33	BTONUM1 142	CPCT 129
ARABIC 18	BUFFON 108	CPDF 129
ARC 151, 178	BWD 24	CRU 25
ARCS 157,177	CAST 22	CTOP 82
AREAS 88	CAST9 22	CTOP1 83
ARG 81	CHBOARD 146	CUBE 79
ASK 167	CHVECT 146	DECFRAC 18
ASTROID 71	CCOF 115	DERANGE 138
AV 103,178	CHISQ 125	DET 31
AVERAGE 162	CI 56	DICE 108
AXISLAB 8,87	CIRC 67	DIF 45
BASE 17	CIRCLES 66,157,177	DIFFSQ 27
BETA 126	COINS 20	DIGSUM 21
BINCOEF 43	COLADD 168	DIST 81
BINFRAC 17	COMBS 140	DIV 24

DRAW 65,155,175
E1 50
E2 50
EHPARM 69
EHCURVE 70
ELLIPSE 67
EPIHYP 75
EST 113
EUC 15
EUL 96
EULER 47
EVAL 17
EVENFAC 14
EXPAND 28
EXPEC 125
EXPRESS 17
EXTEND 76
FACS 13
FACTINT 135
FACTORS 13
FIB 46
FIB1 46
FIB2 47
FILLBOX 150,157,176
FINDALL 172
FPDF 130
FREQ 109
FWD 23
GOFIT 125
GP 33
GPDF 129
GREGORY 47
HAND 145
HANOI 149
HATS 139
HCF 15
HIST 157,178
HYPERB 67
HYPGEOM 101

IF 167
IMAGE 81
INRANGE 112
INT 162
ITEROOT 93
IX 34
JOIN 66,156
KENDALL 119
LAGR 53
LATIN 144
LCM 16
LEXPERM 137
LIFE 148
LIMACON 77
LINES 66,156,176
LNORM 62
LOG 11
LONGEST 61
LOOKUP 10
LOT 168
LOTD 169
MADJ 103
MAGIC 144
MATCHES 102
MATCHPR 102
MAXI 109
MDICE 108
MEDIAN 104
MERGE 163
METRIC 20
MIDPT 97
MM 147
MODE 105
MONOD 42
MONOI 42
MORT 58
MUL 24
MULTAB 9
NEV 54

NEWRAP 94
NEWRAP1 95
NINT 127
NORM 62
NPCT 127
NPDF 126
NUMDER 139
ODDS 7
OFF 54
PADD 32
PARAB 68
PARENV 74
PAS 44
PASN 44
PATTERN 84
PCOFS 36
PCTILE 105
PDIFF 32
PDRAW 114
PERFNOS 14
PERM 137
PERMS 137
PERSPEC 80
PIN 155,175
PINTEG 32
PI1 48
PI2 49
PI3 49
PI4 49
PI5 49
PLOT 65,155,175
PMULT 32
POISSON 100
POLI 54
POLY 32
POLYFIT 113
POWERS 9
PRECTST 23
PRIMES 12
PTOC 83

Index of Programs and Variables

PTS 178	RSORTU 164	SUMSQQ 28
PV 57	RUN 106	SUPSET 35
PY 61	RUNS 117	SVAR 103
PYTH 63	SAMEDIG 21	TAREA 59
QUAD 29	SD 104	TERM 41
RANGE 105	SELECT 141	TERM1 39
RAŃKDWN 162	SEP 19	TERM2 40
RANKUP 162	SER 41	TESTNUM 173
RBO 106	SER1 39	TIMER 174
RECUR 16	SER2 40	TPCT 128
REFLECT 158	SERX 41	TPDF 128
REFLX 68	SIDEA 59	TRANKD 163
REMDUP 162	SIDE3 60	TRANKU 163
REMDUPM 162	SHRTEST 61	TRAP 89
REPLACE 172	SIMPSON 89	TRAPEZ 98
REPLIC 169	SIN 51	TRIGTAB 11
RESID 114	SKETCH 66,156,175	TRIPLE 63
RFD 107	SOMESET 36	TWOWAY 111
RND 106	SORTUP 162	UN 34
RNE 106	SORTDWN 162	UNIT 169
RNO 106	SPEAR 118	UPMATC 164
ROMAN 19	SPROB 121	UPMATR 164
ROOT2 50	SQUTRI 9	UPSORT 165
ROTO 77	SQXXX 9	UPT 168
ROT2 77	SRANK 163	UPTD 169
ROT3 78	SREC2 47	UTEST 123
ROUND 161	SREC4 48	VAR 103
ROWADD 168	SS 115	WITHIN 112
RPT 21	SSIZE 131	WITHOUT 53,170
RPTHATS 139	STEST 120	WTEST 122
RSA 107	STMLEAF 110	XFM 155
RSORTD 164	SUBSET 35	YATES 125

Made in the USA
Coppell, TX
26 January 2024

28205903R00111